The Hands-on XBee Lab Manual

T0296998

The Hands-on XBee
Lab Manual
Experiments that Teach you XBee Wireless Communications

Jonathan A. Titus

AMSTERDAM • BOSTON • HEIDELBERG • LONDON
NEW YORK • OXFORD • PARIS • SAN DIEGO
SAN FRANCISCO • SINGAPORE • SYDNEY • TOKYO
Newnes is an imprint of Elsevier

Newnes is an imprint of Elsevier
The Boulevard, Langford Lane, Kidlington, Oxford OX5 1GB, UK
225 Wyman Street, Waltham, MA 02451, USA

First edition 2012

Notice
No responsibility is assumed by the publisher for any injury and/or damage to persons or property as a matter of products liability, negligence or otherwise, or from any use or operation of any methods, products, instructions or ideas contained in the material herein. Because of rapid advances in the medical sciences, in particular, independent verification of diagnoses and drug dosages should be made

British Library Cataloguing-in-Publication Data
A catalogue record for this book is available from the British Library

Library of Congress Cataloging-in-Publication Data
A catalog record for this book is available from the Library of Congress

ISBN: 978-0-12-391404-0

For information on all Newnes publications
visit our web site at books.elsevier.com

Typeset by MPS Limited, Chennai, India
www.adi-mps.com

Printed and bound in United States of America

12 13 14 15   10 9 8 7 6 5 4 3 2 1

Dedication

I dedicate this book to Jane, my wife and companion for over 40 years, and to our grandchildren, Delia, Samuel, and Laney.

Contents

Foreword to The Hands-on XBee Lab Manual

The late 20th century saw the birth of connectivity. From simple computer networks the Arpanet grew into the Internet. At first the playground of academics, nearly everyone is now connected to the "net" and therefore to everyone else and to vast amounts of data.

The 21st century has seen the dawn of the wireless era. Cell phones are no longer telecomm devices; they're part of the vast Internet. Now we're all connected all of the time, in the office or at the beach. Some of us use a smart phone more for data communications that voice transmissions. Indeed, that mobile device may have four or more antennas: one with a link to satellites in space for navigation, another to the 3G network, a wi-fi connection and Bluetooth for near-field links to headsets and the like.

Radio data communications surround us, from messages sent to the smart sign hanging over the highway to science instruments transmitting their findings from remote Arctic regions to a lab in some pastoral setting via a satellite uplink.

Why have a Bluetooth link from the phone to a headset? A wired approach is a nuisance. It's bulky, in the way, and snags on things. Electronics is cheap; connectors are not, and the wireless version likely saves money and is more reliable. Messing with a tangle of cables in a lab or even with your PC is awkward. It won't be long before those all go away.

It's much more complicated to establish a link over an RF connection than with a wire, but smart hardware costs little today, and canned software is increasingly available. One popular option is Digi International's line of XBee prepackaged radio modules. You don't need to understand the nuances used, like direct-sequence spread spectrum coding or offset quadrature phase-shift keying, because those details are all taken care of by the modules.

You *do* need to understand how to use and interface to the modules, which is not made any easier by the terse and sometimes cryptic manuals. And XBee also uses the old AT command set, which is increasingly hard to find information about. For these reasons Jon Titus's *Hands on XBee Lab Manual* is invaluable.

We learn in different ways. For many of us a hands-on approach is much more efficient than slogging through data books. Jon has taken that approach here, in a series of experiments designed teach by doing. From the very beginning you'll be putting together components that make something neat happen. Early experiments emulate a single-direction wire: a receiver module's output pin mirrors the input pin on the transmitter. Look Ma—no wires!

Each lesson is progressively more complicated and useful. Send analog data through the ether. Control multiple XBee modules. Connect other embedded components, like an Arduino Uno or ARM mbed board to the XBee modules. That, of course, is really the basis of embedded wireless networking.

Explore personal-area networks. These are self-assembling communications links where the network discovers at run time which XBee modules are operating. Jon shows how to do this, and how to piece the network together. Along the way you'll learn to handle interrupts, a crucial concept in the world of embedded systems.

Complex stuff. But fear not: Jon shows every step of each experiment. The lavish illustrations leave no chance for confusion. Whether it's a screen shot of a communications tool or a drawing of how to connect a LED, nothing is left out.

Welcome to the second decade of the 21st century, the age of wireless data communications. This book is your essential guide to using XBee modules to toss off the shackles of wires.

Jack Ganssle

Acknowledgments

Thanks go to the technical-support people at Digi International for their courteous and timely help answering questions and for pointing me to application notes that helped overcome several problems. I also found useful information on Web pages that cater to experimenters and on highly technical Web sites.

I also appreciate the book-proposal reviews from Bradley J. Thompson, Alex Mendelsohn, Ken Gracey, and Dr. Michael Batchelder. All had worthwhile comments and suggestions that improved this book.

Introduction

This book will appeal to engineers, product designers, entrepreneurs, teachers, experimenters, and students who want to learn how to use the popular XBee wireless-communication modules from Digi International without having to master radio engineering or complicated communication protocols. You can power an XBee module and have it communicate with other XBee modules in a few minutes. As always, though, the devil is in the details, so the experiments start with basic information about how to use the Digi International X-CTU software that simplifies the setup and use of XBee modules. You will learn how to use one-way communications from remote modules that report digital and analog information. Experiments also explain how to set up 2-way communications between separate modules or among modules in a small wireless network.

As you learn about communications, you'll gather other information not explained in other books or application notes. That includes using sleep modes, causing a remote module to immediately wake up, how to use pulse-width modulation, and so on. Later experiments introduce the Arduino Uno and ARM mbed microcontroller (MCU) modules and provide code that allows for two-way communications between a PC and a remote MCU module. Instructions explain how to use either an ARM mbed or an Arduino Uno. Software experiments also show how to identify an unknown number of XBee modules in a small network and how to obtain remote digital and analog information and provide it in useful formats. Keep in mind this book does not describe a series of projects, rather it explains how XBee modules communicate digital and analog information, how to understand that information, how to configure modules, how to set up a network of modules, and so on. Then you apply your creativity and design products or build projects based on what you learned here.

When I started to investigate how to use XBee modules I downloaded the technical manual from Digi's Web site and reviewed the specifications and settings. I also looked in forums and technical-support pages to find information about how to use XBee modules. Unfortunately, the technical manual describes a lot of commands and operating conditions, but it doesn't explain how to use that information to do useful things. The forums and tech-support Web sites often addressed specific questions such as "I have a problem getting ABC to work..." or "How do I connect an XBee to a pressure sensor?" That type of information helps only a few people solve a specific problem.

So, I decided to jump into the XBee "experience" to learn what the modules could do and how to use them without prior knowledge. Thus the experiments explain how you can use the XBee modules to communicate with each other and I leave the details of what to communicate up to each reader. Some people might need to transmit analog signals from sensors and others might choose to monitor alarm switches on doors and windows. How you use the modules is up to you, and you will know how to use the XBee modules after you go through the experiments in this book. These experiments do not cover ZigBee-protocol networks. Other books provide that information.

The experiments use standard XBee modules and readily available and inexpensive components and supplies. You can run most of the experiments without any knowledge of electronics or programming, but experience programming with a language such as BASIC or C will help in a few experiments. Experience using digital logic and breadboarding circuits will help, too. You must read schematic diagrams, place components in breadboards, and make connections accordingly. The text explains what to do.

Some explanations might seem elementary, but I find it better to address people unfamiliar with electronics and let experts skip through basic information. Although each unit can serve as a stand-alone experiment, I encourage readers to start with Experiment 1 and go through them sequentially.

I recommend the Digi XBee modules, model number XB24-ACI-001, that operate in the 2.4 GHz Industrial, Medical, and Scientific (ISM) band that requires no license, complies with international standards, and conforms to the Institute of Electrical and Electronic Engineers (IEEE) 802.15.4 standard. You can purchase modules directly from Digi International or from distributors worldwide.

In a few cases, I had an XBee module go off into "unknown territory," probably because of an error in one of my configuration settings. If that happens, you can quickly reset a module with its factory-default settings and get back on track.

The XBee modules can respond to 68 commands, many of which I never used and I doubt most users will need to use. A book of experiments cannot cover every command, so you experiment as you see fit. Because I worked on a lab bench and within a small building I did not change the output radio-frequency power-level (PL) setting from the factory default. Nor did I use any of the seven diagnostic commands.

Before you start the experiments, download and print the latest manual, "XBee/XBee-PRO RF Modules" from the Digi International Web site at: www.digi.com. I had the manual version 90000982_B. This document describes the control commands, the parameters they need, and what they do.

You must have the free Digi X-CTU software that runs on a Windows PC. This software lets you easily monitor XBee activity, send an XBee module command, test communications with a remote XBee module, and configure modem settings. Digi uses the word "modem" for the control portion of an XBee module, but this book simply calls the XBee devices modules. My modules had firmware version 10E6. Digi used hexadecimal values for version numbers and you can check the version number through the X-CTU software.

I ran into a couple of problems with the X-CTU software that deserve mention. First, set up a USB-to-XBee adapter and connect it to your PC via a USB cable before you start the X-CTU software. At times when I started the software first, it did not recognize the USB-to-XBee adapter when I plugged it in. Second, on occasion the X-CTU software quit unexpectedly if I used the mouse trackball or scroll wheel to move up or down within an X-CTU window. If you have set up a long command packet you will lose it. Third, you cannot save packets from within the X-CTU software, so write them down. Finally, if you have boxes open within the X-CTU Modem Configuration window and use a mouse trackball or wheel, the choices in the open box will scroll rather than the larger configuration window.

Now some information specific to the experiments.

CODE DOWNLOADS

You can download all code used in the experiments and use it freely:

http://www.elsevierdirect.com/companion.jsp?ISBN = 9780123914040

Each experiment has a folder that holds the necessary code or XBee-configuration information, although a folder marked "None" indicates an experiment has no corresponding software. To save time, experiment folders include Modem Configuration files, which have the .pro file extension, to use with the X-CTU software to set XBee-module parameters. If you get stuck and something doesn't seem to work, load the provided configuration file into the X-CTU software and save it in the module.

I consider my code open source and hope readers will modify it and share it with others. The code includes an Excel spreadsheet, Packet Creator 2, that will help you create hexadecimal messages or packets used in some experiments or in microcontroller code. You should purchase a calculator that can handle hexadecimal values. Texas Instruments and Casio have several such inexpensive models.

I will post comments, corrections, new code, and other information on the http://www.elsevierdirect.com/companion.jsp?ISBN=9780123914040 site and readers can contact me directly at jontitus@comcast.net. I will do my best to answer questions but cannot guarantee a personal reply. But before you contact me or post a question on a forum, locate helpful XBee-related information on the Internet, on the Digi tech-support pages, and on the support pages for the ARM mbed and Arduino Uno products.

CODING STYLE AND ERROR CHECKING

I have kept programs and flow charts simple so people who do not have much experience can follow their operations. Feel free to change the software to meet your needs. You can always restore XBee modem configurations to factory-fresh settings and it's unlikely you will damage a microcontroller by running programs that contain errors. The C compilers catch syntax errors but not logical errors; those in which you use an incorrect value, perform an incorrect operation, and so on.

In some cases I provide working code that lacks all the steps a professional programmer would include to detect errors and take corresponding actions.

In many cases, error-detection code can take as much room as your "working" code. I'll put a "place holder" in the code and include a note about testing for errors where it seems appropriate. Never assume that your code works properly until you test it thoroughly. And never assume errors cannot occur. I make no claim to special programming expertise and welcome suggestions that help others create better or more-efficient code.

I use a programming "style" that vertically aligns code sections and subsections and find it easier to follow code in this form:

```
byte SerialInput()
{
while(Serial.available() == 0)
    {
    more code here...
    }
return (Serial.read());
}
```

rather than code in the following form:

```
byte SerialInput(){
    while(Serial.available() == 0){
    }
    return (Serial.read());
}
```

where it's difficult to identify pairs of braces that define code sections. Of course, you can use any format you wish.

DEBUGGING CODE

You can do some debugging on an MCU by using an output pin to drive an LED. When the MCU gets to a certain place, it can turn the LED on or off, or flash it at a preset rate. You also can easily track variables and print messages at "checkpoints" in your code. The ARM mbed module provides a separate USB virtual serial port that will send debugging or test information to your PC. For the ARM mbed module I used the Windows HyperTerminal to print such information.

The Arduino Uno module has a USB connection with a host PC, but it shares its serial port with the TX and RX pins on the board, which can cause conflicts unless you do not connect devices to the Uno's serial-port pins. The newer Arduino Mega2560 module and the Digilent chipKIT Uno32 have extra serial ports, but I have not used these modules.

DIGITAL LOGIC NOTATION

The experiments that follow use digital signals that exist in either a logic-0 or a logic-1 state. By convention, engineers use an "over-bar" above the corresponding signal name to indicate a logic-0 level will cause an action to occur, enable another digital device, cause a reset, and so on. Even with a word processor, creating an over-bar causes headaches, so a forward slash (/) in front of a signal name will indicate a logic 0 causes the named action. Thus, a signal named /RESET will reset the device the signal goes to. On the other hand, an input named LOAD, for example, requires a logic 1 to cause a loading action. Some authors use an asterisk in front of signal names in place of a slash.

SCHEMATIC DIAGRAMS OF CIRCUITS

The schematic circuit diagrams in the experiments and appendices use standard electronic symbols for components. Sometimes when people draw circuits and then convert them to a computer-drawn diagram, errors occur. When you see signals come together as shown in Figure I.1a, you might wonder whether they connect at this point or one signal simply passes over the other. To avoid confusion, the experiment circuit diagrams have no 4-point connections. Instead, you see only 3-way connections as shown in Figure I.1b. This drawing style eliminates ambiguity.

INTERMITTENT ARDUINO UNO PROBLEMS

I had intermittent problems with Arduino Uno modules because they use the same serial port at pin 0 (RX) and pin 1 (TX) to connect with external serial devices, such as XBee modules and with the host PC USB connection used

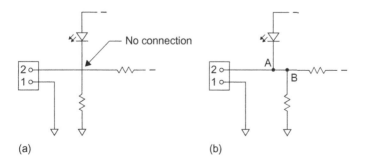

(a) (b)

FIGURE I.1 (a) In this book, the intersection of two lines means a "fly over" of two separate wires. It is *not* a 4-way connection of signals. (b) A dot on the lines indicates when three signals connect at A and at B. To avoid ambiguity, this book uses only three-wire connections in schematic diagrams.

to download programs. Often my Arduino compiler window would display the following information in the bottom text-message window:

Problem uploading to board: See http://www.arduino.cc/en/Guide/ Troubleshooting#upload for suggestions.

Binary sketch size: 2284 bytes...etc...

The Arduino Uno Troubleshooting Guide recommends several solutions, including removing anything connected to pin 0 and pin 1. I found this step annoying and used a switch to disconnect pins Uno 0 and 1 from my external circuits during a code download. I don't know how the Arduino people expect experimenters, hobbyists, and others to continually disconnect and reconnect circuits to the pin 0 and pin 1 contacts. The Arduino Mega and the Digilent chipKIT Uno32 modules provide extra serial pins and do not share them with a PC USB connection.

On occasion, the Arduino Uno compiler would not "find" the attached Uno module and would report finding only the COM1 port. Connecting and reconnecting the Uno module, pressing its reset button and starting and restarting the Uno software did not help.

I also ran into another problem in later experiments that transmit information from the Arduino Uno module to the host PC for display in the Serial Monitor window available within the Uno programming framework. Data transmitted to the Serial Monitor also went to any attached XBee module. This "dual connection" means the XBee module "sees" the data going to the Serial Monitor as a string of bytes that might inadvertently include a valid but unwanted command, or the bytes might cause the XBee module to return an error packet. In several cases, after sending information from the Uno module to the host PC, the XBee module attached to the Uno acted in unexpected ways. The experiments that use a Uno module include steps that help circumvent these problems, but after a short time, disconnecting and reconnecting serial-port lines becomes annoying. Use a switch as described in the relevant experiments.

The Arduino Uno Web site provides information about how to use other I/O pins for serial communications implemented in MCU code rather than in hardware. The limitations of these software-serial communications prevent their use in these experiments with XBee modules. Frankly, if I started writing again I would drop the Arduino Uno and use a newer board that offers several serial ports. Help is at hand.

A NEW 32-BIT UNO BOARD

Shortly after I completed the Arduino Uno experiments, Digilent and Microchip Technology announced the chipKIT Uno32 board that offers software and hardware compatibility with the Arduino Uno module. The chipKIT Uno32 gives you an 80-MHz 32-bit processor (PIC32MX320F128H) that provides a wide variety of I/O devices, 128 kbytes of flash memory, and 16 kbytes of static memory. The MCU includes two UARTs, one for the USB connection with a host PC and one for use with your own devices, such as displays, XBee modules, a small printer, and so on. So you should not have problems with serial-port conflicts. The PIC32MX320F128H MCU connects the extra UART, #2, via the SDI2 (serial-data input) and SDO2 (serial-data output) pins on the Uno32 SPI connector (J8). The MASTER and SLAVE jumpers at JP5 and JP7 let you change the pinouts for the SDO2 and SDI2 signals. Unfortunately, the Digilent schematic diagram for the Uno32 module does not show the difference between the MASTER and SLAVE modes, which simply swap the serial lines on the SPI connector between pins 1 and 4. Pin 6 provides a ground connection.

In time I'll post some code for the chipKIT Uno 32 that shows how to use the SDI2 and SDO2 UART.

LOGIC-LEVEL CONVERSION

Because the XBee modules operate with a +3.3-volt power source they should not connect to 5-volt logic signals. The Appendices include information about how to construct two logic-level-conversion circuits that solve this problem. If you use a 3.3-volt MCU or MCU board such as the ARM mbed, you don't need to convert logic levels.

EQUIPMENT

The Bill of Materials in Appendix F lists the minimum and recommended quantities of components for the experiments. And this appendix includes information about electronic equipment and suppliers.

FREE-SOFTWARE CODE LICENSE

Readers have access to the software included in this book under the following Massachusetts Institute of Technology free-software license:

==========

==========

ACKNOWLEDGMENTS

Thanks go to the technical-support people at Digi International for their courteous and timely help answering questions and for pointing me to application notes that helped overcome several problems. I also found useful information on Web pages that cater to experimenters and on highly technical Web sites.

XBee and XBee-PRO are trademarks of Digi International. Other trademarks belong to their respective owners. I have no financial interest in any of the companies or products mentioned in this book.

Jonathan A. Titus, KZ1G
Herriman, Utah, USA

Introduction to the X-CTU Software

REQUIREMENTS

1 XBee module
1 USB-to-XBee adapter (see text)
1 USB cable—type-A to mini-B
Digi X-CTU software running on a Windows PC, with an open USB port

INTRODUCTION

In this short experiment you will learn how to use the Digi International X-CTU software to program and control XBee modules. These modules use a simple serial communication protocol that connects them to external devices, such as a PC and microcontroller integrated circuits, through a built-in Universal Asynchronous Receiver Transmitter (UART). The UART in the XBee modules operates by default at 9600 bits/second, which makes it compatible with UARTs in other equipment, thus simplifying communications.

Many XBee commands use standard alphanumeric characters in the American Standard Code for Information Interchange (ASCII), which means you can simply type on a PC keyboard to perform XBee-module operations without having to write programs for a computer. Thus early experiments in this book will not require a microcontroller (MCU), although readers familiar with MCUs will quickly understand how to apply them, if they choose. For a table of ASCII characters and values, see Appendix G.

The X-CTU software "connects" with an XBee module via a USB connection that acts like a "virtual" serial port. The software handles the USB drivers needed to make possible these serial communications from a Windows PC. At the XBee end of the USB cable you need an XBee-to-USB adapter such as the XBee USB Adapter Board (part no. 32400) from Parallax or the XBee Explorer USB (part no. WRL-08687) from SparkFun Electronics. Both adapters provide several small LEDs that indicate operating conditions. Refer to

1

the respective adapter data sheets for more information. You must have one of these adapters to perform the experiments in this book.

Step 1. Digi International provides free X-CTU software that you can download from the Digi Web site at: www.digi.com. Search for XCTU (no hyphen) and find the entry, "Knowledge Base Article – X-CTU (XCTU) software" and click on it. On this page, click on the link at the bottom of the page, "Click here to go to the X-CTU download page." Then download the latest version of the X-CTU software and install it. Do not start the X-CTU software.

Step 2. Attach the XBee-to-USB adapter to the USB cable and then attach the USB cable to your PC. If possible, use a USB port on your PC. Extension USB ports on monitors or USB hubs sometimes cause problems. Do not insert an XBee module in the adapter now. You will do so shortly. Follow the adapter manufacturer's instructions (if any) that describe how to load drivers that configure the USB port to act like a serial port. If you plug in the XBee-to-USB adapter after you start the X-CTU software, the software might not detect the adapter.

Step 3. Start the X-CTU software as you would start any other program. The opening display should appear as shown in Figure 1.1. If not already

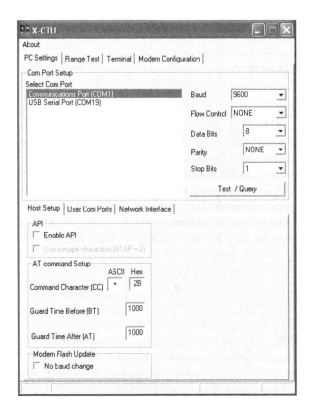

FIGURE 1.1 The X-CTU software opens a window that offers four tabs along the top. To start, click on the PC Settings tab to make a connection between your PC and the XBee-to-USB adapter.

selected, click on the PC Settings tab. In the Serial Com Port window, you should see text similar to:

Communications Port (COM1)

USB Serial Port (COMxx)

where the xx represents a 1- or 2-digit number. Use your mouse to select this line of text, which should highlight it to tell X-CTU to use the COMxx port to connect to the XBee-to-USB adapter. Windows assigns COM-port numbers in sequence and on my lab PC the adapter appeared as COM19. Many late-model PCs lack a serial port, so you might not see the first line shown above for COM1.

The X-CTU software uses the following settings to communicate with an XBee module:

Baud   = 9600 (Equivalent to 9600 bits/second)

Flow Control = NONE

Data Bits   = 8

Parity   = NONE

Stop Bits   = 1

The X-CTU software should start up with these settings, but if not, use your mouse to change them to the settings shown above.

Step 4. Disconnect the USB-XBee adapter board from the end of the USB cable and insert an XBee module into the socket strips on the adapter board. Ensure you have matched the placement of the XBee module with that specified by legends on the adapter, or diagrams in the adapter instructions. Figure 1.2 illustrates possible hardware problems.

After you have properly inserted the XBee module into the USB-to-XBee adapter, reconnect the adapter to your USB cable. Depending on the adapter you use, you might see LEDs turn on or flash. (On the Parallax adapter, a yellow and a green LED turned on and a red LED flashed.)

Step 5. In the X-CTU window again look at the PC Settings section. Click on the PC Settings tab at the top of the X-CTU window if you have clicked on other tabs.

In the PC Settings area, find the button on the right side marked Test/ Query. Click on it to test the electrical communication connection to the XBee module. A "Com test/Query Modem" window will open and it should appear as shown in Figure 1.3.

If you see the error message, "Unable to communicate with modem," shown in Figure 1.4, the X-CTU program cannot "find" the XBee module. Click on Retry to try communications again, or click on OK to go back to the PC Settings window. You also can close the X-CTU program and restart it after you confirm you have the XBee-to-USB adapter properly connected to your PC. The X-CTU program usually does not recognize an adapter board plugged in after you start the software. Also recheck the settings given in Step 3.

Step 6. The information shown earlier in Figure 1.3 indicates this experiment used an XB24 XBee module with firmware version 10E6. Depending on the module you have, model and firmware information might vary from that shown here. The version information uses hexadecimal, or base-16, values.

(a)   (b)

(c)   (d)

FIGURE 1.2 These photos illustrate problems that can occur when you improperly insert an XBee module in an adapter: (a) Incorrect module orientation, (b) a bent pin, (c) incomplete insertion in an adapter or other socket, and (d) a module in an off-by-one-pin position.

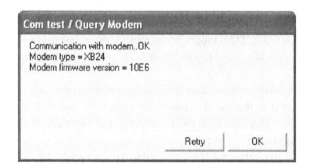

FIGURE 1.3 When the X-CTU software successfully communicates with an XBee module, you will see a display that includes the information shown here.

Step 7. After you have seen the Modem type and Modem firmware version information in the Com test/Query Modem window, click on OK and then select the Terminal tab to open the X-CTU terminal window (Figure 1.5). This window lets you type a message to control the attached XBee module and to see responses from the module. The cursor should already flash in the white message area.

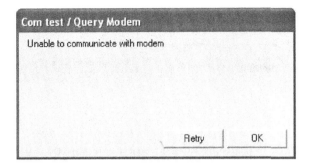

FIGURE 1.4 If you see this screen when you test the connection between the X-CTU software and your XBee module, recheck the settings shown in Steps 3 and 5, ensure you have firm connections at each end of the USB cable, and that you have plugged in the XBee module firmly.

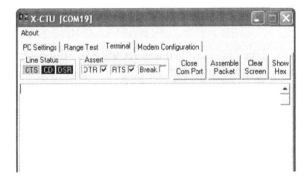

FIGURE 1.5 The top of the X-CTU Terminal provides several settings and buttons. Leave the settings as shown. You will use the Assemble Packet, Clear Screen, and Show Hex button in later experiments.

The XBee modules rely on a set of "AT" commands to set operating conditions and request actions. Years ago engineers created dial-up modems that used similar AT commands to set up modem conditions, initiate communications, dial phone numbers, and so on. Digi International lists the set of XBee AT commands for the XBee24 modules in the document "XBee/XBee-PRO RF Modules" available on the company's web site: http://ftp1.digi.com/support/documentation/90000982_D.pdf as revision D, dated 25 July 2011. Digi might have posted a newer version, though.

An AT command begins with three plus signs, $+++$, sent from your PC. Type $+++$ in the terminal window. DO NOT press Enter or any other key after you type $+++$.

You should see the $+++$ printed in blue and after a few seconds, the letters OK should appear in red at the end of the plus signs (Figure 1.6). The "OK" message lets you know the XBee module can accept AT commands. An XB24 module remains in this "AT-command" mode for about a minute. If you don't

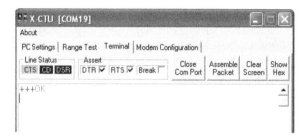

FIGURE 1.6 An OK response from an XBee module that received the characters + + +.

type in a command within that time, you must type + + + again and wait for the XB24 module to again display "OK" in the terminal window.

Step 8. In this step you will put the XBee module into the AT-command mode and then send it a command. Review the command sequence below before you do anything more:

+++

ATVR [Enter]

Now, type the three plus signs and wait for the "OK" response: + + + OK

Then, type ATVR and press Enter. Note what you see in the terminal window. In my lab, I saw:

+++ OK

ATVR

10E6

The ATVR command sends the AT prefix that lets an XBee module know a command will follow. In this case, the VR – Firmware Version command asked the module to reply with the version number for its firmware.

Do not put an XBee module in AT-command mode and type random letters just to see what happens. Doing so could alter internal settings that affect an XBee module's operations.

If you type an invalid AT command or a valid AT command the XBee module cannot perform, it responds with ERROR in red letters below the latest AT command.

Step 9. If you plan to go on to Experiment 2 now, leave your XBee module connected to the USB-XBee adapter, leave the USB cable connected to the adapter and your PC, and do not close the X-CTU window.

Note: For more information about the origin and use of the AT modem commands, visit: en.wikipedia.org/wiki/Hayes_command_set.

How to Change XBee Module Configurations

REQUIREMENTS

1 XBee module
1 USB-to-XBee adapter
1 USB cable—type-A to mini-B
Digi X-CTU software running on a Windows PC, with an open USB port

INTRODUCTION

In this experiment you will learn how to use the X-CTU software to program an XBee module for specific actions. You will make changes in experiments that will follow, so please do not go on to the next experiment until you understand how to change settings and confirm them.

Step 1. You must have an XBee module plugged into a USB-XBee adapter and the adapter must connect to a Windows PC USB port. You also must run the X-CTU software. If you do not have this equipment set up and the X-CTU software running, please complete Experiment 1 before you proceed.

Step 2. Check the connection between your PC and the XBee module: Within the X-CTU window, click on the PC Settings tab and ensure you have the communications set for Baud: 9600, Flow Control: NONE, Data Bits: 8, Parity: NONE, and Stop Bits: 1. Click on the Test/Query button and the Com test/Query Modem window should open and display "Communication with modem..OK" and other information. (If you do not see this message, go back and repeat Experiment 1 and see the Troubleshooting section in Appendix H.)

Click on OK in the message window.

Step 3. Click on the Modem Configuration tab. This window lets you observe and change all of the operating information internal to an XBee module. At this point the configuration window could be blank (white) or it might contain information as shown in Figure 2.1. It does not matter.

7

The Hands-on XBee Lab Manual.

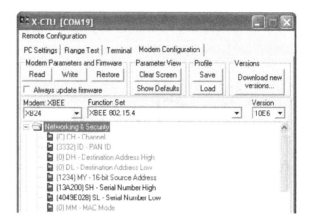

FIGURE 2.1 The Modem Configuration window displays settings from an XB24 XBee module with firmware version 10E6.

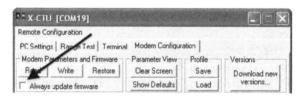

FIGURE 2.2 Leave the "Always update firmware" box unchecked, or uncheck it. You do not need to update firmware and trying to do so could cause problems.

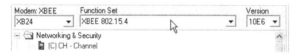

FIGURE 2.3 If not already set as shown here, change the MODEM: XBee and Function Set text boxes to the configurations shown here.

(Digi International often uses "modem" to describe its XBee modules. To me, modem sounds archaic, so I will avoid using it as much as possible.)

Step 4. Before you proceed, uncheck the small box "Always update firmware" on the left side of the Modem Configuration window as shown in Figure 2.2. If you check this box, or leave it checked, the X-CTU software will attempt to update the firmware in an attached XBee module. If that update fails, you could end up with a non-functional module. Do not update firmware.

Step 5. Before you use the X-CTU software to change configuration information, you must ensure the software knows the type of module and the function set you plan to use. When you work with a module, the Modem: XBee should show XB24 and the Function Set should show XBee 802.15.4, as shown in Figure 2.3.

FIGURE 2.4 Modifying an XBee configuration setting changes its color so you can quickly spot modified values or selections.

> Set/read the lower 32 bits of the 64 bit destination address. Set the DH register to zero and DL less than 0xFFFF to transmit using a 16 bit address. 0x000000000000FFFF is the broadcast address for the PAN.
>
> RANGE:0-0xFFFFFFFF

FIGURE 2.5 The text window provides information about a selected module-configuration setting.

The XBee 802.15.4 setting operates the modules according to a standard established for the Institute of Electrical and Electronic Engineers (IEEE) to ensure proper communications between all modules via a fixed protocol.

Step 6. In the Modem Parameters and Firmware box in the upper left section of the X-CTU window, click on Read. This action reads the settings already programmed in your XBee module. The window should now display the parameters in a directory-like format typical of Windows and other operating systems, as shown earlier in the lower part of Figure 2.1.

You should see a folder labeled Networking & Security at the top of the window and below that, settings for CH–Channel, ID–PAN ID, DH–Destination Address High, and so on. Under the Networking & Security heading, move the mouse cursor to DL–Destination Address Low and click on this label. A text box will open to the right of the DL–Destination Address Low label. Disregard any value in this box.

Clear any information in the box and type 4567. The 4567, a hexadecimal value, also appears in the parentheses to the left of the DL–Destination Address Low label, and the label turns olive green (Figure 2.4). That color, which indicates you changed the information, makes it easy to find modified parameters. Unchanged parameters remain bright green, parameters you cannot change appear in black, and errors turn a parameter label red.

A click on a parameter also displays information about the parameter and its allowed settings in the text area at the bottom of the Modem Configuration window (Figure 2.5).

Step 7. Move the cursor to MY–16-Bit Source Address, click on it and in the text box that opens to the right of the label, type 1234. Again, note the X-CTU program placed 1234 in the parentheses at the left of the MY–16-Bit Source Address label.

Step 8. For now, any changes made in the Modem Configuration window exist only within the X-CTU software. An attached XBee module has not yet received them. In the Modem Parameters and Firmware box, click on the Write

```
Getting modem type....OK
Modem's firmware not updated
Setting AT parameters..OK
Write Parameters...Complete
```

FIGURE 2.6 This message shows the X-CTU program successfully transferred the configuration information to an XBee module.

button. Below the list of settings the X-CTU software displays the message, "Getting modem type" and then a bar graph labeled, "Setting AT parameters." When the bar graph disappears, you should see the messages shown in Figure 2.6.

Step 9. These messages indicate the X-CTU program has successfully saved the two parameters you changed in an attached XBee module. You can confirm this action in two ways:

• Click the Read button to obtain the settings from the XBee module you just programmed. To the left of the labels Destination Address Low and 16-Bit Source Address you should see the values typed in earlier.

• In the Modem Configuration window, each label has a 2-letter prefix, such as DL for Destination Address Low and MY for 16-Bit Source Address. These letters represent the AT command used to read or write a parameter value or choice. Click on the Terminal tab to get to the terminal view. In this window, click the Clear Screen button.

To determine the Destination Address Low value, type:

+++

ATDL[Enter]

You should see:

+++OK

ATDL

4567

To determine the 16-Bit Source Address, type:

+++

ATMY[Enter]

You should see:

+++OK

ATMY

1234

Step 10. You also can use the AT commands to set parameters. Just follow the AT command with a new parameter. To change the MY–16-Bit Source Address to 0040, type:

+++

ATMY0040[Enter] (No space between command and a parameter value!)

To query the XBee module, type:

+++

ATMY[Enter]

You should see:

+++OK

ATMY

40

The X-CTU software does not display nor does it need leading zeros in the Modem Configuration window, so the command ATMY0040 and ATMY40 have the same effect. In some cases, described later, commands must include leading zeros, and instructions will explain those situations.

Step 11. To return an XBee module to its factory-default condition, click on the Modem Configuration tab and then in the Modem Parameters and Firmware box, click on Restore. This action resets all parameters to the factory-default settings and automatically writes them into the XBee module's flash memory. You do not have to click the Write button.

To confirm the default settings, click on Read and you will see the labels for all the settings you can modify have returned to green and thus the factory-set values. Before you go to the next experiment, I recommend you restore the default values for the attached XBee module.

Step 12. So far, the XBee module does not transfer information wirelessly. The next experiment takes that step.

Note: Unless specified otherwise, all modem parameters use hexadecimal values. You will find a hexadecimal-binary-decimal converter at: http://www.mathsisfun.com/binary-decimal-hexadecimal-converter.html.

One-Way Digital-Input Communications

REQUIREMENTS

2 XBee modules
2 XBee adapters (see text)
1 USB-to-XBee adapter (see text)
1 USB cable—type-A to mini-B
1 3.3-volt DC power supply (see text)
Solderless breadboard
Digi X-CTU software running on a Windows PC, with an open USB port

INTRODUCTION

This experiment shows you how to configure one XBee module as a wireless transmitter and one as a receiver. You will set the transmitter to continually send information to the receiver, which will display information in the X-CTU Terminal window. The diagram in Figure 3.1 shows this basic arrangement for one-way communication of digital information from the XMTR to the RCVR module. The RCVR sends the received information to the X-CTU Terminal, which converts each byte into a corresponding ASCII character and a hexadecimal value.

Now you will set up a solderless breadboard to hold two XBee modules and supply power to them. The 20-pin XBee modules use male pins spaced 2 mm apart, but typical solderless breadboards space receptacles on 0.1-inch (2.5 mm) centers, so you need adapters that make connections between the pins on an XBee module and the receptacles.

Manufacturers have produced several types of XBee socket adapters. I have used the 22-pin 32403 XBee Adapter Board from Parallax and the 20-pin BOB-08276 Breakout Board for XBee Module from SparkFun Electronics. No matter which adapter you use—or make—keep a diagram handy that shows the pin numbers for the adapter and the XBee-module pins the breadboard receptacles connect to. They do not correspond one-to-one on all types of adapters.

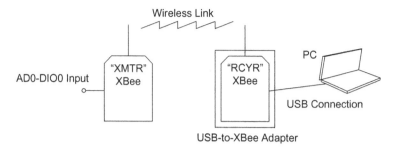

FIGURE 3.1 This experiment configures one XBee module as a transmitter with one active digital input. That module will take two samples, one second apart, and transmit the results to another XBee module labeled RCVR.

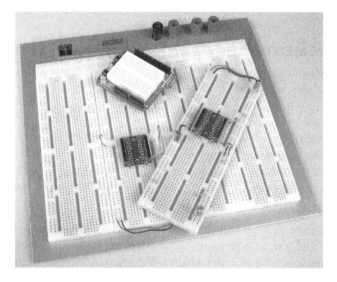

FIGURE 3.2 Breadboards come in a variety of shapes and sizes. Some include marked power buses and terminals.

The experiments in this book always refer to the pin numbers and signal names on an XBee module.

Step 1. In this step, plug two XBee adapters into a solderless breadboard and keep the adapters separated by about 3 inches (7.6 cm). If the breadboard has other components attached to it, please remove them now. Do not put an XBee module in either adapter. In later experiments you will use the second adapter, but it's easier to insert it now when you have nothing else in the breadboard.

Figure 3.2 shows three solderless breadboards. When I make breadboard power connections, I use the *outside* buses for the positive voltage and the *inside* buses for ground. I have used this arrangement for many years because regardless of breadboard orientation, I know where to connect to power and ground. The photo in Figure 3.2 also shows a breadboard, but with color-coded and

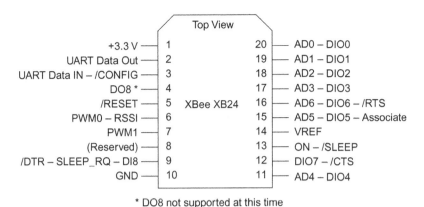

FIGURE 3.3 Pin-and-signal designations for an XB24 XBee module.

labeled power and ground buses. Use any arrangement you like, but I strongly recommend you label or color the ground and the positive-voltage buses unless already labeled. Incorrect power connections can quickly destroy circuits.

I also recommend you label one XBee module XMTR for transmitter and one RCVR for receiver. I used a marker to color the XBee logo green on my transmitter and red on my receiver. In other experiments the XMTR won't always transmit and the RCVR won't always receive, but labels and colors make them easier to keep track of. If you have additional XBee modules, label them, too. (I had two other modules labeled END and PRO.)

Step 2. The diagram in Figure 3.3 shows the pin numbers and signal labels for XBee and XBeePRO modules based on a top-down view. Some pins share functions. The ADx–DIOx pins, for example, can operate as digital inputs or outputs, or as analog inputs for an internal analog-to-digital converter (ADC).

Step 3. As I prepared the experiments for this book I used an Extech Instruments Model 382203 Analog Triple Output DC Power Supply because I had one in my lab. You do not need this type of high-quality power supply for experiments. Although experiments will call for 3.3 volts, two D-size 1.5-volt dry cells in series will suffice to power XBee modules. And batteries work well if you want to locate XBee modules and circuits away from your lab bench. Jameco Electronics sells a 2-cell battery holder (part no. 216390) and DigiKey stocks a similar battery holder (part no. BH2DL-ND).

Connect +3.3-volt power to pin 1 (VCC) on each adapter and connect ground to pin 10 (GND) on each module, as shown in Figure 3.4. Pin numbers *always* refer to XBee pin numbers and *not* to adapter pin numbers due to variation in adapter pin configurations. All pin information refers to a top view of a device.

Step 4. Now you will program the transmitter (XMTR) XBee module. I recommend you always disconnect the USB-to-XBee adapter from its USB cable before you insert or remove an XBee module. This means you have an unpowered USB-to-XBee adapter when you insert or extract a module.

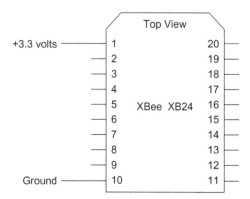

FIGURE 3.4 In this experiment, the XMTR XBee module needs only a power and a ground connection. In a later step you will make another connection.

Disconnect the USB-to-XBee adapter from its cable. Insert your XMTR module in the USB-to-XBee adapter socket and reconnect the adapter to the USB cable.

Step 5. If not already running, start the X-CTU software and confirm it can communicate with the XBee module. Use the PC Settings window or the Terminal to test communications. After you have communications between the module and the X-CTU software, go to the X-CTU Modem Configuration tab and click the Restore button to return configurations to their default condition. Then click Read to obtain those settings.

Step 6. You must change the XBee configurations for the transmitter module to those shown below. If you have problems and cannot get the XBee modules to work properly, download the EX3_XMTR.pro file that provides a ready-to-use configuration for the transmitter. Then in the Modem Configuration window, find the Profile area and click on Load. Locate the appropriate profile and it will load into the Modem Configuration window. Click on Write to save it in an XBee module. Remember, all XBee configurations use either a menu choice or hexadecimal digits.

If you do not use the configuration file, under the Networking & Security heading, change:

DL – Destination Address Low   = 1234
MY – 16-Bit Source Address   = 5678

Each XBee module has a 64-bit (8-byte) serial number as well as a 64-bit destination address divided into a DH and a DL section, each 32 bits, or 4 bytes, long. The DL value sets the 16-bit address of the destination XBee module you want to communicate with. The DH address remains set at 0x0000, where the zero and lowercase x denote a hex value. So, the XMTR module has a destination address of 0x0000 0x1234. The MY value sets the 16-bit source address for the XMTR module. In other words, the MY information gives a module with its own unique 16-bit address.

Step 7. Scroll down in the Modem Configuration window to the I/O Settings heading. In this section clicking on some parameters opens a small text window that lets you select one of several choices. Other parameters require a typed hexadecimal value.

Under the I/O Settings heading, change:

D0 – DIO0 Configuration = 3-DI (Sets data-input for pin 20, AD0/DIO0)
IR – Sample Rate   = 3E8 (Hexadecimal for 1000)
IT – Samples before TX   = 2

Now pin 20 (AD0-DIO0) at the XMTR module will operate as a digital input (DI) pin. The sample rate uses increments of one millisecond, so a value of 0x03E8 (1000_{10}) provides a delay of 1000 milliseconds, or one second, between samples. Once a second the transmitter will sample any XBee input/output pins that you enabled—in this case only AD0-DIO0—and save that information.

The IT parameter set the transmitter to acquire two samples before it transmits data to the receiver. Thus, the receiver should see information from the transmitter once every two seconds.

Step 8. After you have entered the parameters above, recheck them and click on the Write button to save them in the XBee XMTR module attached to the USB-to-XBee adapter.

After you successfully program the XMTR module, you can click the Read button and review the settings to ensure you programmed them properly. If necessary, make changes to the parameters and again click the Write button and then the Read button to confirm the latest changes.

Disconnect the USB-to-XBee adapter from the USB cable and remove the XBee XMTR module. Insert it into the one of the XBee adapters in your breadboard.

Step 9. Now you will program the XBee receiver module, RVCR. Place the XBee module marked as the receiver in the USB-to-XBee adapter and connect the adapter to the USB cable. The X-CTU software should still be running, so confirm communication between the X-CTU program and the RCVR XBee module.

After you have communications between the module and the X-CTU software, go to the X-CTU Modem Configuration tab and click the Restore button to return configurations to their default condition. Then click Read to obtain those settings.

Step 10. You now set parameters in the RCVR module so it can receive information from the XMTR module you just programmed. Click on Read to read the RCVR module's configuration information. As described in Step 6, if you have problems and cannot get the XBee modules to work properly, download the EX3_RCVR.pro file that has the ready-to-use configuration for the receiver.

Step 11. Change the RCVR XBee-module configurations to those shown below.

If you do not use the configuration file, under the Networking & Security heading, set:

DL – Destination Address Low = 5678
MY – 16-Bit Source Address   = 1234

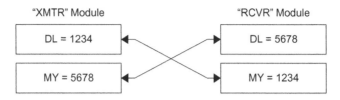

FIGURE 3.5 By setting the MY and DL values as shown here, two XBee modules can communicate with each other.

These two settings relate to those you set for the XMTR module in Step 6. Now you have the RCVR module set with the MY address you used as the destination address in the XMTR module, and the destination address in the RCVR module corresponds to the MY address in the XMTR module. These corresponding addresses, shown in Figure 3.5, will let the modules communicate with each other.

Step 12. Under the I/O Settings heading, look for the IU – I/O Output Enable parameter. Click on it and confirm the RCVR module has it set to 1-ENABLED. If this setting does not equal 1, click on the label and choose the setting 1-ENABLED. This parameter lets a module send data it received from another XBee module to an external device, perhaps an MCU, via the UART. In this experiment, the X-CTU software will receive that information and display it in the Terminal window.

Step 13. After you have changed the DL, MY, and IU parameters shown above, recheck them and click on the Write button to save them in the RCVR XBee module attached to the USB-to-XBee adapter. After you successfully program the RCVR module, click the Read button and review the parameters to ensure you programmed them properly.

Do not remove the RCVR module from the USB-to-XBee adapter. It will receive wireless information from the XMTR module and send it to the X-CTU software through the USB connection.

Step 14. Click on the Terminal tab in X-CTU and click on Clear Screen. Then apply power to the XMTR module. You should see a display of somewhat random characters start to appear in the Terminal window as shown in Figure 3.6.

The X-CTU Terminal does not display messages in text you can read. Instead it sends information in packets of bytes. To display the data as hexadecimal values, click the Show Hex button on the upper right side of the Terminal display. The display now shows the characters on the left side of the screen and the hexadecimal data on the right, as shown in Figure 3.7.

A close look at the hex information shows a repeating pattern:

```
7E 00 0C 83 56 78 3F 00 02 00 01 00 01 00 01 6A

7E 00 0C 83 56 78 3F 00 02 00 01 00 01 00 01 6A

7E 00 0C 83 56 78 3F 00 02 00 01 00 01 00 01 6A

7E 00 0C 83 56 78 30 00 02 00 01 00 01 00 01 7B
```

PC Settings | Range Test Terminal | Modem Configuration |

Line Status ──── ┌ Assert ──────────┐   Close   Assemble Clear Show
CTS CD DSR   DTR ☑ RTS ☑ Break ☐   Com Port   Packet   Screen Hex

```
~. . . .Vx?. . . . . . . .j~. . . .Vx?. . . . . . . .j~. . . .Vx?. . . . . . . .   ▲
.j~. . . .Vx?. . . . . . . .j~. . . .Vx?. . . . . . . .j~. . . .Vx?. . . . .
. . . .j~. . . .Vx?. . . . . . . .j~. . . .Vx@. . . . . . . .i~. . . .Vx?. . .
. . . . . .j~. . . .Vx
<. . . . . . . .m~. . . .VxA. . . . . . . .h~. . . .Vx@. . . . . . . .i~. . .
Vx@. . . . . . . .i~. . . .Vx@. . . . . . . .i~. . . .Vx@. . . . . . . .i~.
. .Vx@. . . . . . . .i~. . . .Vx@. . . . . . . .i~. . . .Vx;. . . . . . . .n|
```

FIGURE 3.6 A raw-data message from an XBee transmitter looks like this.

PC Settings | Range Test Terminal | Modem Configuration |

Line Status ──── ┌ Assert ──────────┐   Close   Assemble Clear Hide
CTS CD DSR   DTR ☑ RTS ☑ Break ☐   Com Port   Packet   Screen Hex

```
~. . . .Vx?. . . . .   7E 00 0C 83 56 78 3F 00 02 00 01 00
. . . .j~. . . .Vx?.   01 00 01 6A 7E 00 0C 83 56 78 3F 00
. . . . . . . .j~. . .   02 00 01 00 01 00 01 6A 7E 00 0C 83
Vx?. . . . . . . .j   56 78 3F 00 02 00 01 00 01 00 01 6A
~. . . .Vx?. . . . .   7E 00 0C 83 56 78 3F 00 02 00 01 00
. . . .j~. . . .Vx?.   01 00 01 6A 7E 00 0C 83 56 78 3F 00
. . . . . . . .j~. . .   02 00 01 00 01 00 01 6A 7E 00 0C 83
Vx?. . . . . . . .j   56 78 3F 00 02 00 01 00 01 00 01 6A
~. . . .Vx?. . . . .   7E 00 0C 83 56 78 3F 00 02 00 01 00
. . . .j~. . . .Vx@.   01 00 01 6A 7E 00 0C 83 56 78 40 00
```

FIGURE 3.7 By splitting the Terminal window into a character column (left) and a hex-values column (right) you can obtain bytes of information that indicate the state of I/O pins at the XMTR module.

Parsing this information according to specifications for XBee communications lets you see what it means. Although the values do not include a 0x prefix they appear next as hexadecimal numbers:

7E   = start of transmission, a preset value (1 byte)

00 0C   = a count of the following bytes in transmission (2 bytes)

83   = code for 16-bit module addressing, preset value (1 byte)

5678   = 16-bit address (2 bytes)

3F   = signal strength of wireless signal at receiver (1 byte)

00   = status byte (1 byte)

02   = number of samples (1 byte)

00 01   = Active-Signal Bytes identify active I/O pins (2 bytes)

00 01   = first of two samples for digital I/O pins (2 bytes)

00 01   = second of two samples for digital I/O pins (2 bytes)

6A   = calculated checksum (1 byte), not included in byte count

Here's how you break down this transmission in more detail:

00 0C	$= 12_{10}$ bytes in the transmission, not counting the checksum, the start-of-transmission byte or these two byte-count values
5678	= the Destination Address Low you programmed into the RCVR module
3F	= the signal strength at the RCVR, now a value of 63_{10}
00	= a status byte
02	= the number of samples you programmed as the Samples-before-TX value in the XMTR module
00 01	= Active-Signal Bytes that identify active analog or digital pins at the transmitter module
00 01	= digital information for active digital inputs (1st sample)
00 01	= digital information for active digital inputs (2nd sample)

Step 15. In the Bit Position row in the following tables, B7 represents the most-significant bit and B0 represents the least-significant bit. Appendix I provides blank tables you can copy for your own use. The information in the two Active Signal Bytes (0x00 and 0x01) goes into Table 3.1 to identify any active I/O pins at the XMTR module. Each I/O pins has a corresponding bit position in these two bytes. The letter X in a Bit-Function position indicates the bit has no use in I/O operations.

The left-most Active-Signal Byte, 0x00, equals 00000000_2, which indicates inactive analog channels A5 through A0 and an inactive digital input at D8. The right-most Active-Signal Byte, 0x01, equals 00000001_2, which indicates inactive digital pins for D7 through D1 (AD7-DIO7 through AD1-DIO1), but an *active* digital pin at D0 (AD0-DIO0). Remember: You set DIO0 as "3 – INPUT" for the transmitter.

Important: The Active-Signal Bytes indicate *which* inputs will have information for the receiving XBee module. They do not indicate the *state* of any active I/O pins. The bytes that follow provide that information.

Table 3.1 Bit Locations in Active-Channel Bytes

First Active-Signal Byte								
	First Hex Character				**Second Hex Character**			
Bit Position	B7	B6	B5	B4	B3	B2	B1	B0
Bit Function	X	A5	A4	A3	A2	A1	A0	D8
Data	0	0	0	0	0	0	0	0
Second Active-Signal Byte								
	First Hex Character				**Second Hex Character**			
Bit Position	B7	B6	B5	B4	B3	B2	B1	B0
Bit Function	D7	D6	D5	D4	D3	D2	D1	D0
Data	0	0	0	0	0	0	0	1

Next you have two values, 0x0001 and 0x0001, that represent the two samples from the XMTR module, taken one second apart. The sampled-data bytes have the arrangement shown in Table 3.2, which includes data from the first pair of bytes, 0x00 and 0x01. Again, an X in a Bit-Function position indicates the bit has no use in I/O operations.

Because the Active-Signal Bytes indicate only the DIO0 pin will have an active signal on it, you can look only at the D0 position in the Second Data Byte. So ignore the D8 through D1 bits in Table 3.2. To re-emphasize; the Active-Signal Byte indicates any active I/O pins and the Data Bytes tell us what signal exists at the corresponding input.

Table 3.2 Byte Arrangements for Digital Inputs

	First Digital-Data Byte							
	First Hex Character				Second Hex Character			
Bit Position	B7	B6	B5	B4	B3	B2	B1	B0
Bit Function	X	X	X	X	X	X	X	D8
Data	0	0	0	0	0	0	0	0
	Second Digital-Data Byte							
	First Hex Character				Second Hex Character			
Bit Position	B7	B6	B5	B4	B3	B2	B1	B0
Bit Function	D7	D6	D5	D4	D3	D2	D1	D0
Data	0	0	0	0	0	0	0	1

Step 16. The data received from the XMTR module includes a checksum byte, 0x7A. The transmitter calculated this value based on the values of the bytes in the message. A receiving XBee module also will calculate a checksum based on the received-message values and will compare it to the received checksum. If the checksum values do not match, a transmission or reception error occurred and the RCVR will "silently" discard the received information. A receiver that detects a checksum error will not send an error message to an attached MCU or other device, so you will usually not know when a checksum error occurs. The receiving module will send an acknowledgement to a transmitting module, however. If the transmitter does not receive an acknowledgement, perhaps due to a checksum error, it will try for as many as three times to complete the communication with the receiver. (The RR command will increase the number of retries, but the experiments in this book use the default value for three.)

The checksum involves only the underlined *message* bytes shown here:

```
7E 00 0C 83 56 78 2F 00 02 00 01 00 01 00 01 7A
```

It ignores the 0x7E start-of-transmission value and the 2-byte byte-count value, 0x000C. To calculate the checksum, add the underlined hexadecimal

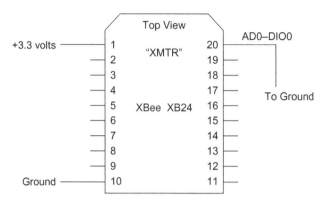

FIGURE 3.8 By connecting the AD0-DIO0 input at pin 20 at the XMTR module, you force this input to a logic-0 state.

values. Here the sum comes to 0x185. Keep only the two right-most (least-significant) hex digits, 0x85, and subtract them from 0xFF. The answer comes to 0x7A.

(For an online hex calculator, visit: http://www.squarebox.co.uk/hcalc. html?0. Appendix E also explains how to download and use an Excel packet-creator spreadsheet that can calculate a checksum value and simplify creating messages used in later experiments.)

Step 17. Although you have only power and ground connected to the XMTR, it transmits a logic 1 to the RCVR as if the D0 input had a logic 1, or about 3 volts connected to it. In an XBee module, internal resistors between the I/O pin and +3.3 volts "pull up" an unconnected input to a logic-1 state. An XBee module has a default condition that turns on these pull-up resistors. In a later experiment you will learn how to use the PR command to turn selected pull-up resistors on or off.

In this step, you will force the D0 input to the logic-0 state to see what happens to the data displayed in the X-CTU Terminal window. Turn off power to your XMTR module. Connect a short wire between the D0 line at pin 20 (AD0-DIO0) and ground (Figure 3.8). This connection pulls the D0 input down to ground, or zero volts, which represents a logic 0. Remember, a logic 0 does not mean "nothing" or an unconnected pin.

Turn on the XMTR module and use the Show Hex view in the Terminal window to observe information received. I observed the following data:

```
7E 00 0C 83 56 78 2D 00 02 00 01 00 00 00 00 7E
```

The Active-Signal Bytes do not change because the AD0-DIO0 pin remains active. But the data at position D0 in the Second Data Byte indicates a logic 0, or ground, at the AD0-DIO0 pin at the XMTR. Use the information for the two data bytes shown in Table 3.2 above to confirm a logic 0 for the D0 line at the XMTR.

You can remove the ground connection at pin 20 on the XMTR module to see if the AD0-DIO0 input goes back to a logic 1.

Step 18. Optional. Go back to Step 7 and use the X-CTU software to configure the AD3-DIO3 pin at the XMTR module as a second digital input pin. You must disconnect the RCVR module and place the XMTR module in the XBee-to-USB adapter. Then return the XMTR and RCVR modules to the breadboard adapter and the XBee-to-USB adapter, respectively. When you power the XMTR module and change the AD3-DIO3 pin (pin 17) from logic 1 (+3.3 volts) to ground (0 volts), you should see a change in the data bytes for the DIO3 pin in the Terminal window. Use the blank tables in appendix I to help parse the new data from the XMTR module.

In the next experiment you will learn how to use the RCVR module to control an external device. You may leave your XMTR and RCVR modules as new set up and use them in Experiment 4.

Use an XBee Module for Remote Control

REQUIREMENTS

2 XBee modules
2 XBee adapters
1 USB-to-XBee adapter
1 USB cable—type-A to mini-B
1 3.3-volt DC power supply
1 LED
1 220-ohm, 1/4-watt resistor, 10% (red-brown-red)
1 Solderless breadboard
Digi X-CTU software running on a Windows PC, with an open USB port

INTRODUCTION

In Experiment 3 you learned how a transmitter module (XMTR) could send information about the state of a digital input to a receiver module (RCVR). The received information shown in the X-CTU Terminal let you determine the logic state of the input pin at the XMTR module. In this experiment you will connect an LED to the RCVR module and control it from the XMTR module. I recommend you perform Experiment 3 before you start this experiment.

Step 1. You will use the same experimental setup used in Experiment 3 for the XMTR module, as shown in Figure 4.1. This figure shows the placement of a jumper wire at the AD0-DIO0 input so you can place a logic-0 or a logic-1 signal on the AD0-DIO0 pin.

Step 2. In this experiment you will configure the RCVR module so its AD0-DIO0 digital output (pin 20) controls an LED. Recall an ADx-DIOx pin can operate as a digital input or output, or as an analog input, depending how you configure it. To set the AD0-DIO0 pin on the RCVR as an output, you must modify the settings in the RCVR module. If you just completed Experiment 3 you should have the RCVR module plugged into the USB-to-XBee adapter and the adapter should connect to a USB port on your PC.

The Hands-on XBee Lab Manual.

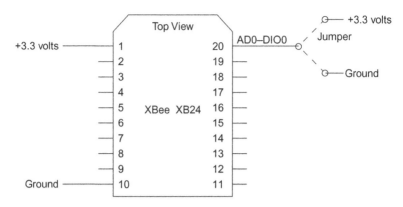

Figure 4.1 Connections for the XMTR XBee adapter in a solderless breadboard for a remote-control experiment.

- If you just completed Experiment 3 and have not made any changes to the RCVR Modem Configuration information, please go to Step 3.
- If you have changed any of the RCVR Modem Configuration information or don't know its configuration, you should reset your RCVR module to its default Modem Configuration. To do so, click on the Restore button and wait for the message: "Restore Defaults..complete" in the bottom X-CTU window. Next, click on Read so you can see the default Modem Configuration settings retrieved from your RCVR module.

If you have problems and cannot get the XBee modules to work properly, download the EX4_RCVR.pro file that provides a ready-to-use configuration for the receiver.

Step 3. Use the X-CTU program to make following four changes to the Modem Configuration settings for the RCVR module:

Networking & Security:
DL – Destination Address Low   = 5678
MY – 16-Bit Source Address   = 1234
I/O Settings:
D0 – DIO0 Configuration   = 5 – D0 High
I/O Line Passing:
IA – I/O Input Address   = FFFF

You must click on the plus sign (+) to the right of the I/O Line Passing folder to open it and see the IA – I/O Input Address configuration. When you move the cursor to, and then click on, the IA – I/O Input Address line, to the right you will see a button labeled Set. Click on it. The Set button opens a text window that shows the I/O Input Address information. Click the Clear button, type in FFFF, and click OK. You should now see the I/O Input Address set to FFFF. This setting lets the receiver accept all I/O-data packets from a transmitter.

The D0 – DIO0 Configuration of "5 – D0 High" places the AD0-DIO0 pin (pin 20) at the receiver module in a "high," or logic-1 state unless a command from a remote module changes its state to a logic 0.

Step 4. Click the Write button to load the new configuration into your RCVR module.

Step 5. If you have not changed or reset the configuration of your XMTR module, please proceed to Step 6. If you have changed, reset, or don't know the settings for your XMTR module, go through the following sequence:

Disconnect the USB-to-XBee adapter from the cable, remove the RCVR module from the adapter, and set it aside. Place the XMTR module in the USB-to-XBee adapter and reconnect it to the USB cable. Open the X-CTU Modem Configuration window and click the Restore button. After you see the message "Restore Defaults..complete" click the Read button. Now change the following configurations shown below for the XMTR module:

DL – Destination Address Low	= 1234
MY – 16-Bit Source Address	= 5678
D0 – DIO0 Configuration	= 3-DI
IT – Samples before TX	= 02
IR – Sample Rate	= 3E8

After you recheck these settings, click on Write to load them into your XMTR module. Disconnect the USB-to-XBee adapter from the cable and remove the XMTR module from the adapter. Return the XMTR module to its adapter socket in your breadboard. Replace the RCVR module in the USB-to-XBee adapter and reconnect it to the USB cable.

If you have problems and cannot get the XBee modules to work properly, download the EX4_XMTR.pro file that provides a ready-to-use configuration for the transmitter.

Step 6. The XMTR module requires a wire on the breadboard between pin 20 (AD0-DIO0) and +3.3 volts, as shown earlier in Figure 4.1. This wire provides a logic-1 or a logic-0 input for the DIO0 input pin. If your XMTR module does not have this connection, add it now.

Click on the Terminal tab in the X-CTU window, clear it, and then apply power to your XMTR module. You should see a string of bytes, such as that shown below, represented as hexadecimal values:

```
7E 00 0C 83 56 78 30 00 02 00 01 00 01 00 01 79
```

As you learned in Experiment 3, these bytes represent:

7E	= start of transmission, a preset value (1 byte)
00 0C	= a count of the following bytes in transmission (2 bytes)
83	= code for 16-bit module addressing, preset value (1 byte)
5678	= 16-bit address (2 bytes)
30	= signal strength of wireless signal at receiver (1 byte)
00	= status byte
02	= number of samples (1 byte)
00 01	= Active-Signal Bytes identify active I/O pins (2 bytes)
00 01	= first of two samples for digital I/O pins (2 bytes)
00 01	= second of two samples for digital I/O pins (2 bytes)
79	= calculated checksum (1 byte), not included in byte count

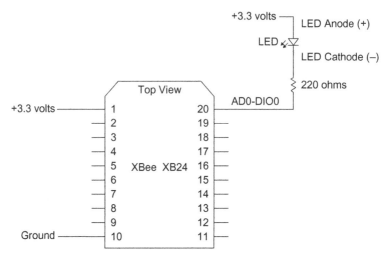

Figure 4.2 This diagram shows power and ground as well as an LED and resistor connected to an XBee adapter socket.

The two samples indicate a logic-1 signal at the DIO0 input on the XMTR module. Your signal strength and checksum values will likely differ from those shown here.

Step 7. Turn off power to the XMTR module. If you have not plugged a second XBee adapter into the solderless breadboard, choose an open section about 3 inches (7.5 cm) away from the XMTR module. Place the second XBee adapter so it orients the XBee socket in the same direction as the adapter already in use for the XMTR module. That means both adapters should have XBee pins 1 through 10 face the same side of your solderless breadboard. This orientation helps avoid confusion over pin numbers and locations. Insert the second, empty XBee adapter in your breadboard.

The second XBee adapter requires +3.3 volts connected to pin 1 (VCC) and ground connected to pin 10 as shown on the left side of the XBee module in Figure 4.2. Make these connections now.

Step 8. On your solderless breadboard, connect one end of a 220-ohm resistor (red-red-brown) to pin 20, AD0-DIO0, on the *unoccupied* XBee adapter. Connect the other end to an unused column of contacts on your breadboard. This column should have nothing else connected to it.

Connect a light-emitting diode (LED) between the free end of the 220-ohm resistor and power. An LED provides two ways to identify its proper connection. The anode (+) has the longer of the two leads. Most individual LEDs also have a flat side on their retaining ring, as shown in Figure 4.3. The flat side indicates the cathode (−) lead.

Next you will connect the LED's anode (+) lead to the +3.3-volt bus on your breadboard and then connect the cathode (−) lead to the free end of the 220-ohm resistor in the otherwise-unconnected column of contacts. When the

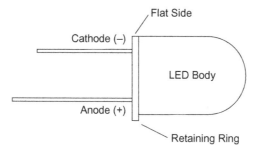

Figure 4.3 Electrical-lead identification for a discrete LED.

DIO0 output of an XBee module in this adapter becomes a logic 0, or ground, current will flow through the LED and resistor and into the DIO0 pin and the LED will turn on. When the DIO0 pin becomes a logic 1, it produces a voltage close to 3.3 volts, so no current can flow and the LED will turn off.

Step 9. Disconnect the USB-to-XBee adapter from the USB cable and remove the RCVR module from the adapter. Ensure you have turned off the 3.3-volt supply to your breadboard. Check the RCVR module for proper orientation with the XBee outline on the unoccupied adapter board and plug in the receiver module.

Step 10. Apply power to your breadboard to turn on the XMTR and RCVR modules. Note the state of the LED. Is the LED on or off?

If you have connected pin 20 (AD0-DIO0) on the XMTR to +3.3 volts, the LED should not light. Move the connection on pin 20 (AD0-DIO0) on the XMTR from +3.3 volts to ground. What happens now? The LED should turn on, but perhaps not immediately.

Remember, the XMTR transmits new information only every two seconds. It has a delay of one second between samples and it takes two samples, so it can take as long as two seconds for the LED to change state after you change the jumper at the XMTR from +3.3 volts to ground, or vice versa. If you think a digital input will change faster than every two seconds, you might need to increase the sample rate at the transmitter accordingly so an XBee receiver module gets the digital information in a timely fashion.

You now have wireless control of the LED at the RCVR module based on the logic input at the DIO0 pin at the XMTR module. You could use a switch, pushbutton, or other mechanical device instead of the jumper at the XMTR module to cause an action at the RCVR module. The XBee modules operate from a 3.3-volt power supply, so you also could connect digital inputs to 3.3-volt-compatible logic devices such as a microcontroller or individual logic circuits. Also, you could use several digital inputs and outputs in parallel.

In the next experiment, you will learn how the XBee modules can transfer analog signals--voltages between ground (0 volts) and +3.3 volts.

Step 11. Optional. If you would like the LED to respond faster, what could you do?

You could reduce the time between transmissions on the XMTR module and also go from two samples to only one sample per transmission.

Program the XMTR for a sample rate of 1/4 second and only one sample. Remember 1/4 second equals 250 milliseconds, for a setting:

IR – Sample Rate   = FA
IT – Samples before TX   = 1

Or you could use the IC – DIO Change Detect command to configure the XMTR module to transmit new digital-input information as soon as the DIO0 input changes state from a logic 0 to a logic 1 or vice versa. You will learn how to do that in another experiment.

XBee Modules Transfer and Control Analog Signals

REQUIREMENTS

2 XBee modules
2 XBee adapters
1 3.3V DC power supply
1 LED
1 220-ohm, 1/4-watt resistor, 10% (red-red-brown)
1 10-kohm (10,000-ohm) variable resistor (trimmer)
1 USB-to-XBee adapter
1 USB cable—type-A to mini-B
Solderless breadboard
Digi X-CTU software running on a Windows PC, with an open USB port
Small screwdriver, flat blade

INTRODUCTION

In Experiment 4 you learned how an XBee transmitter (XMTR) can send digital (on/off) information to an XBee receiver module (RCVR). XBee modules also can communicate voltage information. When you change the voltage on a pin at a transmitter module an output on a receiver will change its output accordingly. Each XBee XB24 module provides two voltage outputs and six voltage inputs. A digital signal exists in only one of two discrete states; logic 0 or logic 1. An analog voltage can exist at any voltage, say 1.236 volts or 1.249 volts, without any defined voltage increments or steps.

Step 1. You will use an experimental setup that includes two XBee adapters in a solderless breadboard. As shown in Figure 5.1, each adapter should have only power (+3.3 volts at pin 1) and ground (0 volts at pin 10) connected to it. If you have other connections, please remove them now. Turn off power to the breadboard.

31

The Hands-on XBee Lab Manual.

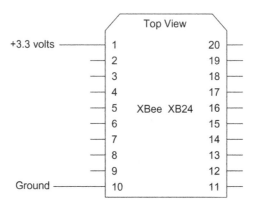

FIGURE 5.1 For this experiment you need two XBee adapters inserted in your solderless breadboard and each adapter must connect to power and ground. Pin numbers refer to XBee-module pins.

Ensure you have the X-CTU program running on your Windows PC. If you have the transmitter (XMTR) or receiver (RCVR) module in an adapter on the breadboard, remove the modules and set them aside. Leave the adapters in the breadboard. For this experiment, both modules require programming with new parameters.

Step 2. Six of the general-purpose I/O pins, AD0-DIO0 through AD5-DIO5, will accept an analog voltage. An internal analog-to-digital converter (ADC) will convert the voltage to a 10-bit binary value. That value will range from 0x000 to 0x3FF. The connection of the reference-voltage input (pin 14, VREF) on a module to a stable voltage establishes the upper limit of the voltage signal the ADC can convert to a digital value. The ADC always has a lower limit of 0 volts. In practice you would use a stable voltage-reference integrated circuit to provide a reference for ADC measurements. Experiments will use the +3.3-volt power, which will provide acceptable results. The VREF input must not exceed the +3.3-volt supply voltage and it cannot go below 2.08 volts.

If you supply a VREF equal to 3.3 volts, the maximum ADC output, 0x3FF, corresponds to a 3.3-volt signal into the ADC. If you supply a VREF equal to 2.5 volts—a common reference voltage—then the maximum ADC output, 0x3FF, corresponds to a 2.5-volt signal into the ADC.

Step 3. In this step, you will use a variable resistor, also called a potentiometer, or "pot," to change the voltage applied to a pin on a transmitter XBee module. We often call these small variable resistors "trimmers," or "trim pots" because electronic designs include them to slightly adjust, or trim, a voltage. Figure 5.2 shows several types of trimmers suitable for use in a solderless breadboard. In most cases, the middle terminal connects to an internal wiper that moves across a fixed resistor that connects to the two end terminals. The resistance between the end terminals does not vary.

FIGURE 5.2 Trimmer potentiometers, or simply "trimmers," come in many package types suitable for use in a breadboard.

Trimmer manufacturers use a numeric code to indicate the resistance between the two end terminals. You will find some 10-kohm trimmers marked "10K" and others marked "103." The 103 means a 10 followed by 3 zeros, thus: 10,000 ohms. Likewise, a 502 trimmer would have a resistance of 5000 ohms, or 5 kohms; 50 followed by two zeros. Some trimmers have one turn—actually about 270-degrees—and other types offer 10 turns in a larger housing. The 10-turn trimmers offer greater accuracy. These experiments use an inexpensive 1-turn trimmer.

To determine the proper connections on a 1-turn trimmer either: (a) Download a data sheet from the supplier, or (b) Hold the trimmer and turn the pointer fully clockwise. The arrow points to one of the end terminals. Now turn the pointer fully counterclockwise and the arrow points to the other end terminal. The middle terminal connects to a wiper—shown as the arrow in a schematic circuit diagram—that provides the varying resistance.

Place a 10-kohm trimmer resistor on the breadboard and ensure each pin goes in its own column of contacts on the breadboard. Then wire it as shown in Figure 5.3. Connect one of the trimmer's end terminals to ground and connect the other end terminal to +3.3 volts. Connect the trimmer's wiper terminal to the adapter pin that goes to pin 20 (AD0-DIO0) for the XMTR XBee module. Also, connect XBee module pin 14 (VREF) to the +3.3-volt power bus.

Turning the trimmer from one extreme to the other varies the voltage at the wiper terminal—and thus at the AD0-DIO0 input—between 0 and +3.3 volts. The XMTR module will convert this voltage to a 10-bit value and transmit it to the RCVR module.

Step 4. Place the XMTR module in the USB-to-XBee adapter.

Click the Restore button in the X-CTU window. After you see the message "Restore Defaults..complete" click the Read button. Now set the configurations as follows. If you wish, you can load this configuration from the file EX5_XMTR.pro:

Networking and Security:
DL – Destination Address Low = 1234
MY – 16-Bit Source Address = 5678

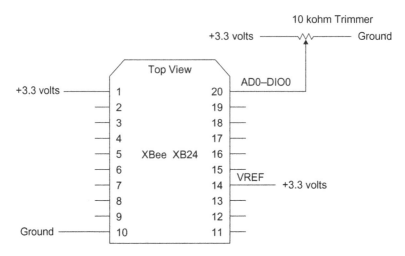

FIGURE 5.3 Wiring diagram for the XMTR XBee adapter in a solderless breadboard. Pin numbers refer to XBee-module pins.

I/O Settings:
D0 – DIO0 Configuration   choose: 2-ADC
IT – Samples before TX   = 02
IR – Sample Rate   = 03E8

Instead of setting the AD0-DIO0 line (pin 20) as a digital input, you have now set it to provide an analog input for the ADC in the XMTR module.

Click on Write to save this configuration information in the XMTR module. Insert the XMTR module into the adapter with the 10-kohm trimmer connected to it.

Step 5. Place the RCVR module in the USB-to-XBee adapter.

Click the Restore button. After you see the message "Restore Defaults.. complete" click the Read button. Now set the configurations as follow. If you wish, you can load this configuration from the file EX5_RCVR.pro:

Networking & Security:
DL – Destination Address Low = 5678
MY – 16-Bit Source Address   = 1234

Under the "I/O Line Passing" heading, go to the line labeled "P0-PWM0 Configuration" and click on it. Change this setting to:

P0 – PWM0 Configuration   2-PWM OUTPUT
and change:
IA – I/O Input Address   = FFFF

The 2-PWM OUTPUT setting configures pin 6 (PWM0-RSSI) as the output at the RCVR XBee module that corresponds to the AD0-DIO0 voltage input on the XMTR module.

Step 6. Click on Write to save this configuration in the RCVR, but leave the RCVR module in the USB-to-XBee socket. You will use the X-CTU program to monitor the information sent by the XMTR module.

Step 7. Switch the X-CTU display to the Terminal screen, clear it, and change to the Show Hex setting if the Terminal window is not already in this mode.

Apply power to the XMTR module and use a small screwdriver to slowly adjust the potentiometer. Remember, the XMTR module takes two readings one second apart and then transmits them to the RCVR module, so data will appear and show changes only every two seconds.

In the Terminal window you should see data that looks somewhat like:

```
7E 00 0C 83 56 78 2E 00 02 02 00 02 41 02 41 F6

.....

7E 00 0C 83 56 78 2E 00 02 02 00 00 00 00 00 26
```

To stop receiving data just turn off the XMTR module. You can turn it on again later. The following breakdown shows the structure of the received information. The upper portion duplicates information seen previously in Experiment 3. The lower portion also might look somewhat familiar.

7E	= start of transmission
00 0C	= number of bytes in transmission
83	= code for 16-bit module addressing
5678	= 16-bit address
2E	= signal strength
00	= status byte
02	= number of samples
02 00	= Active-Signal Bytes
02 41	= first of two samples
02 41	= second of two samples
F6	= checksum

The information in Table 5.1 includes data from the first Active-Signal Byte, 0x02 (00000010_2), and the second Active-Signal Byte, 0x00 (00000000_2), to indicate an active AD0-DIO0 pin as an analog input and no active digital I/O lines.

Because you have no active digital I/O pins on the XMTR module, it will not transmit any digital information to the RCVR module. The XMTR module will send data only for *active* inputs. So in this experiment you will see two 10-bit samples in each transmission from the XMTR because you set it to take two samples from one ADC input and transmit them.

The two ADC values appeared in the transmission as 0x0241 and 0x0241 on my PC. In this example they show equal values because I didn't adjust the trimmer while the XMTR performed the analog-to-digital conversions. (You may go ahead and adjust the trimmer to see how it affects your ADC values.)

The information in Table 5.2 shows how to interpret the two bytes that hold a 10-bit value from the transmitter's ADC. The two most-significant bits in the

Table 5.1 The Active-Signal Bytes Sent by a Remote XBee Module let you Determine which Analog Inputs and Digital I/O Pins are Active

	First Active-Signal Byte							
	First Hex Character				Second Hex Character			
Bit Position	B7	B6	B5	B4	B3	B2	B1	B0
Bit Function	X	A5	A4	A3	A2	A1	A0	D8
Data	0	0	0	0	0	0	1	0
	Second Active-Signal Byte							
	First Hex Character				Second Hex Character			
Bit Position	B7	B6	B5	B4	B3	B2	B1	B0
Bit Function	D7	D6	D5	D4	D3	D2	D1	D0
Data	0	0	0	0	0	0	0	0

Table 5.2 Arrangement of Bits Received from a 10-Bit Analog-To-Digital Conversion at a Remote XBee Module

	First Analog-Data Byte							
	First Hex Character				Second Hex Character			
Bit Position	B7	B6	B5	B4	B3	B2	B1	B0
Bit Function	X	X	X	X	X	X	A9	A8
Data	0	0	0	0	0	0	1	0
	Second Analog-Data Byte							
	First Hex Character				Second Hex Character			
Bit Position	B7	B6	B5	B4	B3	B2	B1	B0
Bit Function	A7	A6	A5	A4	A3	A2	A1	A0
Data	0	1	0	0	0	0	0	1

10-bit value arrive in the First Analog-Data Byte, 0x02 (00000010_2), and correspond to bits AD9 and AD8. The remaining eight bits from the ADC arrive in the Second Analog-Data Byte, 0x41 (01000001_2), and correspond to bits AD7 through AD0. Keep in mind these labels refer to bits in the 10-bit ADC output, not to I/O pins AD0, AD1, and so on. The XBee-module pins have complete labels, such as AD2-DIO2, and so on.

Step 8. What voltage does the 0x0241 value represent? You can combine the 10 bits to 1001000001_2, or keep it as 0x241. When you convert 0x0241 to decimal you get 577. The maximum value for a 10-bit converter equals 0x3FF, or 1023, but the results include 1024 possible values. The ratio of 577 to 1024 multiplied by the VREF voltage (+3.3 volts) gives you the voltage applied via the trimmer to the AD0-DIO0 input at the XMTR module.

You transferred this value via a wireless link, and:

(577 / 1024) * 3.3 volts = 1.86 volts

So, the 1.86-volt "signal" now appears as the hex value 0x0241 at the receiving XBee module. So at the time of my first measurements, the trimmer supplied 1.86 volts to pin 20 on my XMTR. How would you interpret the 10-bit ADC value 0x16F for a 2.5-volt reference? (Here's where a hexadecimal calculator can help convert values.)

0x16F = 367

(367 / 1024) * 2.5 volts = 0.90 volts

Remember, you can only perform a conversion when you set an input pin for an ADC input, and only AD0-DIO0 through AD5-DIO5 have this ADC capability.

Although we know the voltage at my trimmer as a number, it would help if the receiver module could actually produce an analog voltage. It can, as you'll learn in the next steps.

Step 9. Turn off power to your breadboard. Remove the RCVR module from the USB-to-XBee adapter and insert it into the second (unused) XBee adapter on your breadboard.

Connect one lead of a 220-ohm resistor (red-red-brown) to the XBee RCVR module PWM0-RSSI signal at pin 6 as shown in Figure 5.4. Insert the other resistor lead into an unused column of contacts on your breadboard. Connect an LED between the resistor lead in the unused column of contacts and +3.3 volts. Connect the LED's long lead (anode, +) to +3.3 volts. Connect the LED's shorter lead (cathode, −) to the resistor lead in the unused column of contacts.

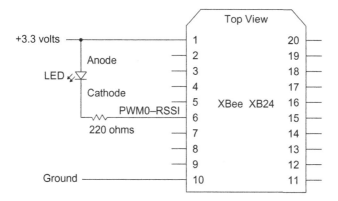

FIGURE 5.4 This diagram shows the connection of power and ground to the RCVR XBee module as well as an added LED and resistor connected between +3.3 volts and the PWM0-RSSI signal at pin 6.

Step 10. Set the trimmer at the XMTR module about halfway between its end positions. Turn on power to your breadboard. The LED connected to the RCVR module should turn on within a second or two.

Adjust the trimmer in small increments. Remember, it can take as long as two seconds for the XMTR module to send updated voltage information to the RCVR module. Do you remember why? You set the XMTR module to take two samples one second apart and then transmit the values from the ADC, so you might not see an immediate change in LED brightness when you vary the trimmer setting. As you slowly adjust the trimmer between its end points you should see the LED go from off to full brightness.

Step 11. Optional: If you would like a faster response, reprogram the XMTR module with a shorter delay between samples:

IR – Sample Rate = 0x03E8 (1-second delay)

For a 1/4-second sample rate, use 250 milliseconds (0xFA) or for a 1/10th-second sample rate, use 100 milliseconds (0x64).

Remember to first Read the parameters from the XMTR module, then modify the IR – Sample Rate parameter and save it in the XMTR module. Place the module back in its socket adapter and turn on power. Does the LED respond faster to trimmer changes? It should.

Step 12. In this step you will learn how the voltage output works on the RCVR module.

It appears the RCVR module produces a voltage between 0 and +3.3 volts to match the voltage measured at the XMTR module. The changing voltage would account for the changes in the LED's brightness. But the RCVR does not operate this way. Instead, the PWM0 output at pin 6 produces a pulse-width-modulated (PWM) output that comprises a series of electrical pulses at a constant frequency. The received voltage value determines the *width* of these pulses.

The plot in Figure 5.5 shows voltage (y axis, 1 volt/division) vs. time (x axis, 50μsec/devision). Most of the time the PWM0 output remains at 0 volts, so current flows from +3.3 volts through the LED, which turns on. Although our eyes cannot see the LED turn on or off, it does so, and the long LED-on periods mean the LED appears bright. In Figures 5.5, 5.6, and 5.7, the horizontal line labeled T represents the trigger-voltage level of 750 mV.

Figure 5.6 shows another plot of voltage vs. time for the PWM0 output but with the trimmer at the XMTR module positioned about halfway between its end points. Now the PWM0 output at the RCVR module is on or off for about half the time. The LED appears at about half brightness because it is on (logic 0) for a shorter time than that shown in Figure 5.5.

The pulses shown in Figure 5.7 cause the LED to turn on only during the short logic-0 periods, so it appears dim, or perhaps off. The frequency of the pulses—about 16 kilohertz, or 16 kHz—remained the same in each of the three plots. Only the pulse widths changed.

Step 13. In some cases you need a real voltage output, not a series of pulses. You can build an electrical low-pass filter that smoothes the pulses into a voltage output. The circuit diagram in Figure 5.8 shows a simple filter that

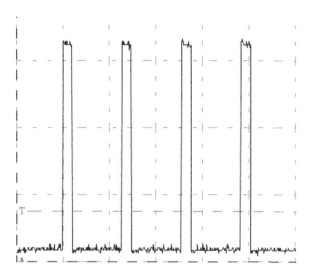

FIGURE 5.5 This pulse-width-modulated output shows an output in a logic-0 state, or ground, most of the time. You would see this type of output when you have set the trimmer on the XMTR module close to 0 volts.

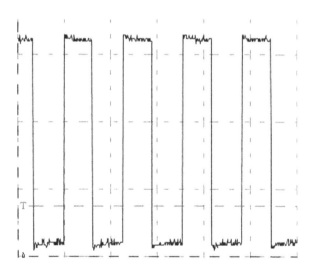

FIGURE 5.6 A PWM output in which the on and off times are approximately equal for each pulse period. The trimmer on the XMTR module is set about halfway between 0 and +3.3 volts.

averages the high and low parts of the PWM signal to create a steady voltage. This filter circuit replaces the LED circuit shown earlier in Figure 5.4.

Depending on whether you changed the sample period for the XMTR module, it still might take a second or two for the voltage output to reflect a change

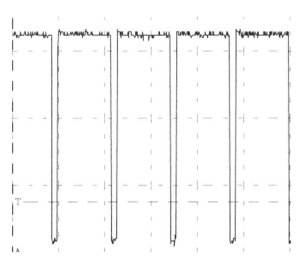

FIGURE 5.7 The pulses in this plot of voltage vs. time remain at 3.3 volts most of the time. Thus, the trimmer at the XMTR module is set close to its 3.3-volt end.

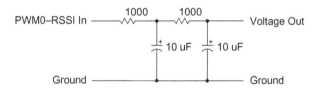

FIGURE 5.8 This simple low-pass filter produces an average voltage output from the pulses created at the PWM0 output at the RCVR XBee module.

at the XMTR trimmer because the XMTR still samples its ADC twice before it transmits the values to the RCVR module.

An XBee module has two independent PWM outputs; pin 6 (PWM0-RSSI) and pin 7 (PWM1). Digi International has associated each PWM output with a specific ADC input. The AD0-DIO0 analog input controls the output at a receiver's PWM0-RSSI pin (pin 6) and the AD1-DIO1 input controls the receiver's PWM1 pin (pin 7).

If you must create an output on PWM0, configure the PWM0 output on the receiver for PWM, *and* configure the AD0-DIO0 input as an ADC input on the transmitter.

Step 14. The RCVR module does not "know" the reference voltage (VREF) used by the ADCs on the XMTR module. It simply passes along the ADC values, which range from 0x000 to 0x3FF. As explained earlier in Step 2, if you supply a VREF equal to 3.3 volts, the maximum ADC output, 0x3FF, corresponds to a 3.3-volt signal into the ADC. If you supply a VREF equal to 2.5 volts, then the maximum ADC output, 0x3FF, corresponds to a 2.5-volt signal into the ADC.

So, if you have a 2.5-volt reference voltage at the XTMR module, the ADC provides the value 0x3FF for a measured voltage of 2.5 volts and 0x000 for a measured voltage of 0 volts, or ground.

But at the RCVR module, the PWM output continues to produce signals that vary from 0 to 3.3 volts as their widths change. If you use a filter circuit, such as that shown in Figure 5.8, on the PWM output you see the filter's Voltage Output signal vary from 0 volts to 3.3 volts rather than from 0 to 2.5 volts measured at the transmitter. You have two ways to handle this mismatch:

1. Use precision resistors to create a voltage divider that scales the filtered PWM output to the same range provided by the ADC at the transmitting XBee module. Then use an operational-amplifier (op amp) follower circuit as a high-impedance-input, low-impedance-output buffer. The simple circuit shown in Figure 5.9 uses two resistors to produce a 0-to-2.5-volt output from a 0-to-3.3-volt input, assuming a 2.5-volt VREF input at the XMTR module. The Microchip Technology MCP6004 op amp serves as the buffer.

I wanted a maximum of 2.5 volts at the + input to the op amp and chose the 4750-ohm resistor as a starting value. Next I used Ohm's law to calculate the current through the 4750-ohm resistor for a 2.5-volt signal:

$I = E / R$

So, 2.5 volts across this resistance yields:

$2.5 \text{ V} / 4750 \text{ ohms} = 0.526 \times 10^{-3} \text{ amperes}$

That small current—about half a milliamp—will not exceed the specification for the PWM0 output on the RCVR module. The current through the other two resistors from the 3.3-volt input also must equal 0.532×10^{-3} amperes and the voltage across those resistors equals: 3.3 volts − 2.5 volts, or 0.8 volts.

Using Ohm's law again:

$I = E / R \text{ or } R = E / I$

$0.8 \text{ volts} / 0.526 \times 10^{-3} \text{ amperes} = 1521 \text{ ohms}$

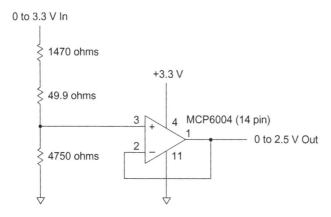

FIGURE 5.9 This circuit uses a voltage divider to scale a 0-to-3.3-volt signal to a 0-to-2.5-volt output. An op-amp buffer provides the output signal.

You cannot find a 1521-ohm resistor, but a 1470-ohm resistor in series with a 49.9-ohm resistor will give you 1519.9 ohms, which comes close enough. These resistors all have a 1-percent tolerance. Although the XBee ADC provides a resolution of one part in 1024, or about 0.1 percent, in practice, you will get about 1-percent accuracy, which matches well with the 1-percent resistor tolerance.

Op-amp circuits go beyond the scope of this series of experiments, though. Find useful information in the following references:

*   Baker, Bonnie, "Using Single Supply Operational Amplifiers in Embedded Systems," AN682, Microchip Technology, http://ww1.microchip.com/downloads/en/AppNotes/00682D.pdf.
*   "Op Amps for Everyone," Texas Instruments document number: SLOD006B. http://focus.ti.com/lit/an/slod006b/slod006b.pdf.
*   "A Single-Supply Op-Amp Circuit Collection," Texas Instruments document number: SLOA058. http://www.ti.com/lit/an/sloa058/sloa058.pdf

2. Process the *data* from an ADC at a remote module with software that runs on a small MCU. As you saw in the first part of this experiment, the RCVR module provides the ADC data in the hex values it sends out its serial port. Later experiments explain how you can connect an XBee receiver to an MCU, parse packets, and then calculate voltages based on the known reference voltage at a transmitter.

Notes

If you wish, you can set the PWM0 or PWM1 output as a received signal-strength indicator (RSSI). You would select RSSI for the PWM0 Configuration or for the PWM1 Configuration in the X-CTU software via the Modem Configuration window. Then connect an LED as shown in Figure 5.4 to indicate the signal strength.

If you set the PWM0-RSSI output (pin 6) for RSSI on the RCVR module and place your hand over the XMTR or RCVR module you can affect the signal strength measured at the receiver. You should see RSSI output vary as indicated by the brightness of the attached LED. It's unlikely you will need an RSSI indication except for testing and debugging an electronic system based on XBee modules.

The article, "PWM RSSI signal – ATRP command for analog RSSI indication" on the Digi International Web site provides a circuit for a 3-level signal-strength indicator. See: http://www.digi.com/support/kbase/kbaseresultdetl.jsp?id=2031.

Remote Control of Digital and Analog Devices

REQUIREMENTS

2 XBee modules
2 XBee adapters
1 3.3V DC power supply
2 LEDs
2 220-ohm, 1/4-watt resistors, 10% (red-red-brown)
1 4700-ohm, 1/4-watt resistor, 10% (yellow-violet-red)
1 10-kohm, 1/4-watt resistor, 10% (brown-black-orange)
1 10-kohm (10,000-ohm) variable resistor
1 Solderless breadboard
1 USB-to-XBee adapter
1 USB cable—type-A to mini-B
Digi X-CTU software running on a Windows PC, with an open USB port
Small screwdriver, flat blade

Optional
1 39-kohm, 1/4-watt resistor, 10% (orange-white-orange)
1 Photoresistor, PDV-P9003-1 (see Bill of Materials in Appendix F)
1 5V DC power supply

INTRODUCTION

In Experiments 4 and 5 you learned how an XBee transmitter can remotely control a digital or an analog output at a receiver. In this experiment, you will configure a transmitter to send both analog and digital information. The receiver will drive two LEDs, one with a digital (on/off) output and one with

The Hands-on XBee Lab Manual.

an analog output that varies LED brightness. You also will learn how to parse the digital and analog information received from a transmitter that has active digital and analog inputs and takes several samples.

Step 1. If you have a powered breadboarded circuit, please turn off the power. Remove the XMTR module from its socket adapter, place it in the USB-to-XBee adapter and connect the adapter to the PC's USB cable. Run the X-CTU program and test communications with the XMTR module.

If you don't know the state of configuration settings, click on Restore and then click Read to obtain the default settings.

Now set the configurations as follows:

DL – Destination Address Low	= 1234
MY – 16-Bit Source Address	= 5678
D0 – DIO0 Configuration	= 2-ADC
D3 – DIO3 Configuration	= 3-DI
IT – Samples before TX	= 02
IR – Sample Rate	= 3E8

You can find this configuration profile in EX6_XMTR_A.pro.

Recheck these settings and save them in the XMTR module. The parameter settings above duplicate those from Experiment 5 but now include a setting to configure pin 17 (AD3-DIO3) as a digital input.

Step 2. Place the XMTR module in its adapter socket on the breadboard. Ensure you have the power (pin 1) and ground (pin 10) connections to the XMTR module. Add a jumper wire between +3.3 volts and pin 17 on the XMTR XBee module. This wire lets you change the logic state of the DIO3 input on the XMTR module as shown in Figure 6.1. If not already connected, add the 10-kohm trimmer and the reference-voltage (VREF) connections.

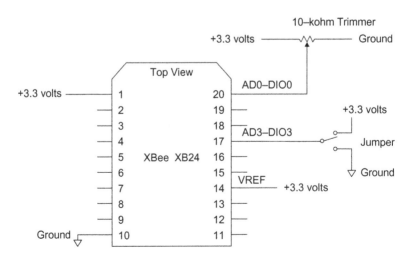

FIGURE 6.1 This XMTR module has a digital input at its AD3-DIO3 pin and an analog input at its AD0-DIO0 input.

Step 3. Insert the RCVR module in the USB-to-XBee adapter and reconnect this adapter to the USB cable. If you don't know the state of configuration settings, click on Restore and then click Read to obtain the default settings.

Now set the configurations as follow:

DL – Destination Address Low = 5678
MY – 16-Bit Source Address   = 1234
D3 – DIO3 Configuration   = 5 D0-HIGH
I/O Line Passing:
P0 – PWM0 Configuration   = 2-PWM Output
IA – I/O Input Address   = FFFF

You can find this configuration profile in EX6_RCVR_A.pro.

Recheck these settings and save them in the RCVR module. Leave the RCVR module in the USB-to-XBee adapter connected to the PC. The parameter settings above duplicate those from Experiment 5 but add a setting to configure pin 17 (AD3-DIO3) on the RCVR module to output a digital signal.

Step 4. Set the trimmer connected to the XMTR module at about halfway between its end points. Change the X-CTU program to the Terminal window and set it to display hex values. Clear the Terminal window. Turn on power to the breadboard and watch as the X-CTU program displays hex values. When I ran this experiment, I found these results; your results will vary:

```
7E 00 10 83 56 78 2F 00 02 02 08 00 08 02 02 00 08 02 02 5B

7E 00 10 83 56 78 2F 00 02 02 08 00 08 02 02 00 08 02 02 5B

.....
```

In this experiment you should see more bytes of information than you saw in Experiments 4 and 5, in which you had *either* a digital *or* an analog input at the XMTR module. Now you have a digital input *and* an analog input.

You can break down the hexadecimal message shown above into two sections:

7E   = start of transmission
0010   = number of bytes in transmission
83   = code for 16-bit module addressing
5678   = 16-bit address
2F   = signal strength
00   = status byte
02   = number of samples
0208   = Active-Signal Bytes
0008   = first digital-input sample
0202   = first analog-input sample
0008   = second digital-input sample
0202   = second analog-input sample
5B   = checksum

Take the two Active-Signal Bytes 0x02 and 0x08 and fill in the spaces in Table 6.1 with the equivalent binary bits. Can you tell which inputs are active at the transmitter?

Table 6.1 This Table lets you Determine which Analog or Digital Pins the Transmitting Module has in use

	First Active-Signal Byte							
	First Hex Character				Second Hex Character			
Bit Position	B7	B6	B5	B4	B3	B2	B1	B0
Bit Function	X	A5	A4	A3	A2	A1	A0	D8
Data	0							
	Second Active-Signal Byte							
	First Hex Character				Second Hex Character			
Bit Position	B7	B6	B5	B4	B3	B2	B1	B0
Bit Function	D7	D6	D5	D4	D3	D2	D1	D0
Data								

The first Active-Signal Byte—0x02 or 00000010_2—indicates you have an active A0 analog-input channel and the second Active-Signal Byte—0x08 or 00001000_2—indicates you have an active D3 digital-input channel, which corresponds to the AD3-DIO3 input (pin 17). Those settings should match the configuration you programmed in the XMTR module.

When an XBee module transmits *both* digital and analog information in two samples, the received digital and analog information for sample one appears first, followed by digital and analog information for sample two, and so on. Given the first pair of Digital-Data Bytes, 0x00 and 0x08, fill in the spaces in Table 6.2 to determine the state of the any active digital inputs identified in Table 6.1. Remember to look only at the digital information for the *active* inputs.

Now look at the value in the first sample received from the ADC at the AD0-DIO0 input. The first pair of bytes from the ADC equals 0x02 0x02, or 00000010_2 00000010_2. Table 6.3 shows the arrangement of bits for an ADC. Fill in the bits in Table 6.3 to determine the value from the transmitter's ADC.

As explained in Experiment 5, you can convert the ADC data, 0x0202 in my results, to a decimal value and compute the voltage applied at the AD0-DIO0 pin (pin 20) at the XMTR. I calculated a voltage of +1.7 volts for my values. Your experimental setup will probably have different analog values, depending on how you set the trimmer in your XMTR-module circuit. Calculate the voltage for your 10-bit ADC value displayed in the X-CTU terminal window. For a hexadecimal calculator, visit: http://www.squarebox.co.uk/hcalc.html?0.

Step 5. In this step, clear the Terminal window, and turn on the XMTR module. Watch the digital and analog data change in the Terminal window as you move the wire connected to pin 17, the AD3-DIO3 digital input, to ground, then to +3.3 volts, then back to ground. You should see the first Digital-Data Byte value remain 0x00. The second byte will change from 0x00 for a logic-0 (ground) input to 0x08 for a logic-1 (+3.3 volts) input at the AD3-DIO3 digital input.

Table 6.2 This Table lets you Determine the State of all Active Digital Inputs Identified by the Active-Signal Bytes in a Transmission from a Remote XBee Module

| | First Digital-Data Byte | | | | | | | |
	First Hex Character				Second Hex Character			
Bit Position	B7	B6	B5	B4	B3	B2	B1	B0
Bit Function	X	X	X	X	X	X	X	D8
Data	0	0	0	0	0	0	0	
	Second Digital-Data Byte							
	First Hex Character				Second Hex Character			
Bit Position	B7	B6	B5	B4	B3	B2	B1	B0
Bit Function	D7	D6	D5	D4	D3	D2	D1	D0
Data								

Table 6.3 Fill in this Table to show the Bits in the two Bytes Received from an Active ADC input in a Transmission from a Remote XBee Module

| | First Analog-Data Byte | | | | | | | |
	First Hex Character				Second Hex Character			
Bit Position	B7	B6	B5	B4	B3	B2	B1	B0
Bit Function	X	X	X	X	X	X	A9	A8
Data	0	0	0	0	0	0		
	Second Analog-Data Byte							
	First Hex Character				Second Hex Character			
Bit Position	B7	B6	B5	B4	B3	B2	B1	B0
Bit Function	A7	A6	A5	A4	A3	A2	A1	A0
Data								

Also change the trimmer settings to confirm you can see the 10-bit ADC value change. You can turn off the XMTR's power to stop transmissions so you can read the digital and analog information. Remember, the XMTR still acquires two samples from the digital and ADC inputs before it transmits, so you should see a new transmission of two sets of data every two seconds.

Step 6. Turn off power to your breadboard. In this step, you will configure another analog input at the XMTR to see how it affects the received data.

Remove the RCVR module from the USB-to-XBee adapter and set it aside. Remove the XMTR module from its adapter on the breadboard and insert it into the USB-to-XBee adapter. Change to the Modem Configuration window and read the configuration data from the XMTR module.

Leave all Modem Configuration information as it is for the XMTR module, but set:

D1 – DIO1 Configuration   = 2-ADC

You can find the complete configuration profile in EX6_XMTR_B.pro. The RCVR module needs no configuration changes.

Recheck your configuration settings and save them in the XMTR module. Remove the XMTR module from the USB-to-XBee adapter and insert it in its breadboard adapter socket. Connect the AD1-DIO1 pin (pin 19) on the XMTR module to the two resistors as shown in Figure 6.2 to create a constant voltage for this analog-input pin.

Place the RCVR module back in the USB-to-XBee adapter. Clear the X-CTU Terminal window and turn on power to the breadboard. Here's the data I observed with the input-data bytes underlined:

```
7E 00 14 83 56 78 3F 00 02 06 08 00 00 02 3E 01 4E 00 00 02 3E

01 4D 42
```

Again, turn off power to the XMTR module to stop transmissions so you can read the information in the Terminal window. Use Tables 6.1, 6.2, and 6.3 to decode the information above. Then decode the information shown in your Terminal window. Find blank tables you can copy in Appendix I. I put the answers for my series of data bytes at the end of this experiment.

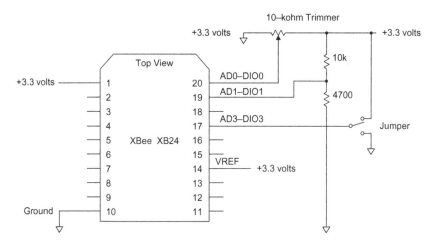

FIGURE 6.2 The circuit for this part of Experiment 6 requires these connections to create two analog inputs (pins 19 and 20) and a digital input (pin 17).

Step 7. In this step, you will connect two LEDs to the RCVR module and control one as a digital (on/off) device and the other as a variable (analog) device. You'll increase the sample rate so you can see changes at the LEDs occur faster.

Turn off power to your breadboard. Remove the XMTR module from the breadboard and set it aside. Remove the RCVR XBee module from the USB-to-XBee adapter and set it aside, too. Remove the wire at the XMTR adapter between pin 19 and the 10-kohm and 4700-ohm resistors. Figure 6.1 shows the proper wiring for the XMTR module. Place the XMTR module in the USB-to-XBee adapter.

Set the XMTR-module configurations to:

DL – Destination Address Low   = 1234
MY – 16-Bit Source Address   = 5678
D0 – DIO0 Configuration   = 2 – ADC
D1 – DIO1 Configuration   = 0 – DISABLED
D3 – DIO3 Configuration   = 3 – DI
IT – Samples before TX   = 01
IR – Sample Rate   = 64

You can find the complete configuration profile in EX6_XMTR_C.pro. Recheck the XMTR configuration information and save it in the XMTR module. Place the XMTR module in its adapter on the breadboard.

You should have a second XBee adapter in your breadboard. If you do not, insert it now. Next add the components described in the next section and shown in Figure 6.3 to the RCVR module.

* Connect a 220-ohm (red-red-brown) resistor between +3.3 volts and an open column on your breadboard. Connect an LED between the open end of the 220-ohm resistor and the RCVR pin 6 (PWM0-RSSI). Ensure you

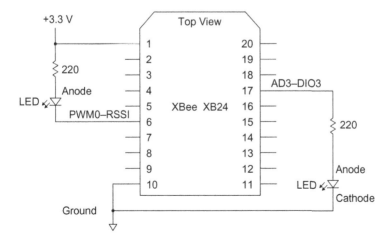

FIGURE 6.3 The RCVR module should have two LEDs connected to it, one to the PWM0 output and another to the DIO3 output.

connect the longer LED lead to the 220-ohm resistor and the shorter lead to the PWM0-RSSI pin.

- Connect a 220-ohm (red-red-brown) resistor between RCVR pin 17 (AD3-DIO3) and an open breadboard column. Then connect an LED between the free end of the 220-ohm resistor and ground. Ensure you connect the shorter LED lead to ground and the longer lead to the 220-ohm resistor.

Set the RCVR-module parameters to:

DL – Destination Address Low = 0x5678
MY – 16-Bit Source Address = 0x1234
D3 – DIO3 Configuration = 5 D0-HIGH
I/O Line Passing:
P0 – PWM0 Configuration = 2-PWM OUTPUT
IA – I/O Input Address = 0xFFFF

You can find this configuration profile in EX6_RCVR_C.pro.

Place the RCVR module in its adapter on the breadboard.

Step 8. Apply power to your breadboard. The brightness of the LED connected to RCVR pin 6 (PWM0-RSSI) should vary from off to full-on as you change the trimmer setting at the XMTR module. As you move the wire that connects to XMTR pin 17 (AD3-DIO3) between ground and +3.3 volts, and back to ground, the LED connected to pin 17 (AD3-DIO3) at the RCVR module should turn on or off.

You have successfully controlled an analog and a digital device at the RCVR module based on an analog voltage and a digital signal at the XMTR module. The XMTR and RCVR modules could be placed 100 feet or more from each other and still perform this type of control.

Important: The digital I/O pins on XBee modules have a one-to-one relationship. If, for example, you set AD4-DIO4 as a digital input on a transmitter and set AD4-DIO4 as an output on a receiver with the corresponding DL and MY addresses, you have direct control of the output at the receiver by changing the state of the input pin on the transmitter. The circuit does not need additional components. Because an XBee module has as many as seven ADx – DIOx pins, you can independently control as many as seven remote devices. You also could employ seven remote modules and control one digital output on each one from one transmitter.

OPTIONAL STEPS

The next steps substitute a cadmium-sulfide (CdS) photoresistor and a fixed-value resistor for the trimmer potentiometer. This circuit will let you control the brightness of the LED on the PWM0 output at the RCVR module based on the light intensity at the photoresistor.

Step 9. Turn off power to your breadboard and remove the 10K-ohm trimmer connected to the XMTR module. Now insert the photoresistor between the ground wire that went to the trimmer and the wire that connects to pin 20

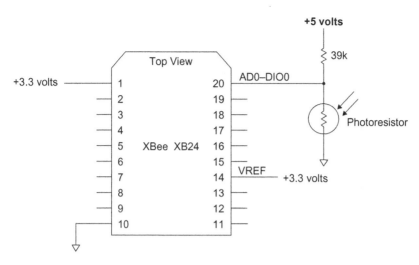

FIGURE 6.4 In this circuit, a photoresistor and a fixed-value resistor form a voltage divider that produces about a 3-volt output in complete darkness to about a 0.1-volt output in bright light.

(AD0-DIO0) at the XMTR module. The photoresistor does not have a polarity, so you do not have to think about a cathode or an anode. Insert the 39-kohm resistor between the +3.3-volt wire that went to the trimmer and pin 20. Figure 6.4 shows a circuit diagram for the photoresistor and its connection to the XMTR module.

Step 10. Turn on power to the breadboard and shine a light on the photoresistor. You should see the LED connected to pin 6 (PWM0-RSSI) on the RCVR module decrease in brightness. As you block the light to the photoresistor, the LED should decrease in brightness. *This part of the experiment uses a 5-volt power supply for the photoresistor circuit* so the circuit provides a wide range of voltages to the XMTR module's ADC input.

The PDV-P9003 photoresistor has a log-log response which means you do not see a simple one-to-one ratio between illumination on the photoresistor and light output from the LED. Download the Advanced Photonix data sheet for a graph of photoresistor resistance versus illuminance in units of lux. This photoresistor will respond to light between violet and red in the visible spectrum and it has the maximum sensitivity at 520 nanometers (nm), which corresponds to green light.

Step 11. Turn off power to your breadboard and remove the RCVR module from its adapter. Insert the RCVR module in the USB-to-XBee adapter. Change the X-CTU window to the Terminal view and clear the Terminal window. Illuminate the photoresistor as brightly as you can and turn on power to the breadboard. Look in the Terminal window for the hex values that correspond to the ADC data from the XMTR module. What do you see? (You can turn off power to the XMTR module after you see several packets of information from the XMTR so you can better read the ADC values.)

With a 60-watt lamp placed a few inches above the photoresistor, I read the hex value 0x000F in the Terminal window; the value received from the ADC on the XMTR module.

Now repeat the measurement with the photoresistor darkened. Do you see a change?

I saw ADC values between 0x3EE and 0x3F5, which is almost a full-scale reading (0x3FF) from the ADC.

You could use other resistive sensors in place of the photoresistor and fixed resistor. Use Ohm's Law and choose resistance values so the sensor output has a range between 0 volts and the VREF voltage applied to the XBee module VREF input at pin 14. The VREF input must not exceed +3.3 volts.

ANSWERS TO QUESTIONS IN STEP 6

For the string of hex values I obtained:

```
7E 00 14 83 56 78 3F 00 02 06 08 00 00 02 3E 01 4E 00 00 02 3E
01 4D 42
```

I "decoded" the following:

02   = number of samples
0608 = indicates active analog and digital channels:
        AD0-DIO0 active as an analog input
        AD1-DIO1 active as an analog input
        AD3-DIO3 active as a digital input

First sample:
0000 = AD3-DIO3 input at a logic 0
023E = AD0-DIO0 analog input (574_{10})
014E = AD1-DIO1 analog input (334_{10})

Second sample:
0000 = AD3-DIO3 input at a logic 0
023E = AD0-DIO0 analog input (574_{10})
014E = AD1-DIO1 analog input (334_{10})

Your measurements for the DIO3 and ADC0 inputs will vary from mine, depending on how you set the trimmer and whether you have the AD3-DIO3 input connected to +3.3 volts or ground. The 0x014E reading from ADC1 represents 1.07 volts, which corresponds to the 1.05-volt value calculated for the 10-kohm and 4700-ohm resistor voltage-divider circuit.

How to Transmit Data on a Logic-Level Change

REQUIREMENTS

2 XBee modules
2 XBee adapters
1 3.3-volt DC power supply
1 10-kohm (10,000-ohm) variable resistor
1 Solderless breadboard
1 USB-to-XBee adapter
1 USB cable—type-A to mini-B
X-CTU software running on a Windows PC, with an open USB port
Small screwdriver, flat blade
Copies of the tables from Appendix I

INTRODUCTION

In Experiment 6, an XBee module transmitted data continuously based on the Sample-Rate and the Samples-before-TX parameters set with the X-CTU program. If an equipment designer selected a long sample rate—a long period between samples—the equipment could miss detecting a *change* at one of the XBee module's digital inputs when it occurs. Suppose you have connected sensors to the digital inputs on an XBee transmitter to detect when certain doors or windows open while homeowners work outdoors. You want an XBee receiver to let them know immediately about a window that opens, not, say, 10 minutes later when the XBee transmitter takes its periodic sample of window and door sensors. In this experiment you will learn how to configure a transmitter so a change on one or more of its digital inputs forces it to transmit digital-input information.

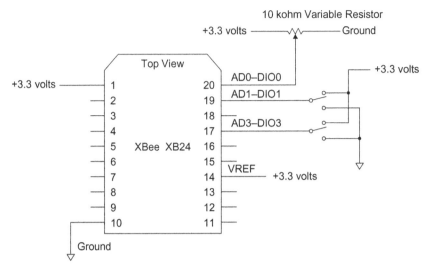

FIGURE 7.1 The connections in this circuit let you change digital inputs from logic-0 to logic-1 states, or *vice versa*, and immediately transmit digital data.

Step 1. In this experiment the XMTR XBee module will have two digital inputs and one analog input. Before you change any connections, turn off power to the breadboard circuits. The circuit diagram in Figure 7.1 shows the needed connections.

These connections to the variable resistor provides an analog signal for the AD0-DIO0 input (pin 20). The jumpers provide digital signals at the AD1-DIO1 (pin 19) and the AD3-DIO3 (pin 17) inputs. The connection between VREF (pin 14) and +3.3 volts supplies a reference voltage for the analog-to-digital converter (ADC) in the XMTR module.

Remove the XMTR module from its adapter, place it in the USB-to-XBee adapter, and connect the adapter to the PC's USB cable. Run the X-CTU program and test communications with the XMTR module.

In the Modem Configuration window, click on Read to obtain the configuration from the XMTR module. If you don't know the state of configuration settings, click on Restore and then click Read to obtain the default settings. In the X-CTU Modem Configuration window, set the XMTR-module configurations as shown below. Do not save these settings in the XMTR module until instructions tell you to.

DL – Destination Address Low = 1234
MY – 16-Bit Source Address = 5678
D0 – DIO0 Configuration = 2-ADC
D1 – DIO1 Configuration = 3-DI
D3 – DIO3 Configuration = 3-DI
IT – Samples before TX = 0A; 10_{10} samples
IR – Sample Rate = 3E8; 1-second sample interval

Step 2. Next you will create a value for the IC-DIO Change-Detect parameter, located under the I/O Settings heading. This hexadecimal value lets you select one or more digital inputs, DIO7 through DIO0, that can cause an immediate transmission of digital-input information. When a *selected input* changes its state—from a logic 1 to a logic 0, or *vice versa*—it forces the XMTR module to transmit the logic state of each active digital input. The IC-DIO Change-Detect parameter comprises one byte, as shown in Table 7.1.

Table 7.1 IC-Change-Detect Settings for Inputs DIO7 Through DIO0

| | IC- DIO Change Detect | | | | | | | |
	First Hex Character				Second Hex Character			
Bit Position	B7	B6	B5	B4	B3	B2	B1	B0
Bit Function	DIO7	DIO6	DIO5	DIO4	DIO3	DIO2	DIO1	DIO0
Data	0	0	0	0	1	0	0	0

For this experiment, assume a *change* in the logic state at the AD3-DIO3 (pin 17) input will force the XMTR module to transmit data. You need a 1 at position DIO3 in the IC-Change-Detect byte to correspond with an input at AD3-DIO3. Place a 0 in the other seven bit positions so you have 00001000_2, which translates to 0x08.

Within the Modem Parameters window, set:

IC – DIO Change Detect   = 08

for the XMTR module. Find this configuration profile in EX7_XMTR_A.pro.

Review the settings and save them in the XMTR module. The XMTR will now gather 10 measurements (IT = 0x0A) one second apart (IR = 0x3E8) before it transmits all samples to the RCVR module. Thus new information will appear in the X-CTU Terminal window every 10 seconds and the messages will hold 10 samples of digital- and analog-input data.

I chose the setting of ten samples for two reasons: First, so you can see how to break down a transmission to locate all ten samples. And second, so you will have sufficient time between the transmissions scheduled for every 10 seconds to force a transmission of digital data.

Step 3. You must understand how an XBee transmitter works when you use the IC-Change Detect settings:

• If an XBee module has gone part-way through a sampling sequence, it will transmit any samples it has already acquired followed by the digital-input information transmitted in response to a digital-input change.

• A forced transmission of digital-input information will not include information from active ADC inputs. You will get only the digital-input information.

• You can force a transmission of digital-input states only when the XBee module detects a logic-level change at a digital input selected with a corresponding 1 in the IC-Change Detect command byte.

- A logic-1-to-logic-0 or a logic-0-to-logic-1 transition will trigger a transmission for a change on an input selected with the IC-Change Detect parameter for the transmitter.

Step 4. Place the XMTR module in its adapter on the breadboard. Check the connections to the XMTR module and the breadboard.

Step 5. Place the RCVR module in the USB-to-XBee adapter and reconnect this adapter to the USB cable. If you don't know the state of configuration settings, click on Restore and then click Read to obtain the default settings.

Set the RCVR-module parameters to:

DL – Destination Address Low = 5678
MY – 16-Bit Source Address = 1234
D3 – DIO3 Configuration = 5 D0-HIGH
P0 – PWM0 Configuration = 2-PWM Output
IA – I/O Input Address = FFFF

Confirm these settings and save them in the RCVR module. Find this configuration profile in EX7_RCVR_A.pro. Leave the RCVR module in the USB-to-XBee adapter connected to the PC. You will use the X-CTU program to monitor data from the XMTR module.

Step 6. Set the variable resistor connected to the XMTR module at about halfway between its end stops. Switch the X-CTU program to the Terminal window and set it to display hex values. Turn on power to the breadboard and watch as the X-CTU program displays data in hex characters. It should take 10 seconds for the first transmission to start. After you receive one or two transmissions of 10 samples, turn off power to the breadboard. You need only one set of information to analyze.

When I ran this experiment, I found the following hex values in the Terminal window. Reformatted values makes it easier to understand them. Your data will look a bit different:

```
7E 00 30 83 56 78 2B 00 0A 02 0A

00 0A 01 D7

00 0A 01 D7

00 0A 01 D7

00 0A 01 D7

00 0A 01 D7

00 0A 01 D8

00 0A 01 D8

00 0A 01 D8

00 0A 01 D8

00 0A 01 D7
```

Here's how this hex information breaks down:

7E   = start of transmission
0030   = number of bytes in transmission
83   = code for 16-bit module addressing
5678   = 16-bit address
2B   = signal strength
00   = status byte
0A   = number of samples
020A   = Active-Signal Bytes

Use copies of the tables in Appendix I to decode the Active-Signal Bytes 0x020A value and determine which analog or digital inputs are active at the XMTR module.

You should find the AD0-DIO0 (pin 20) pin configured as an active analog input and the AD3-DIO3 (pin 17) and AD1-DIO1 (pin 19) pins set for use as digital I/O signals.

Again use the tables from the Appendix to decode one of the 10 samples of analog- and digital-input information in the four columns of hex values above. The first two bytes, 0x00 and 0x0A, provide digital data from AD3-DIO3 and AD1-DIO1. The last two bytes, 0x01 and 0xD7, provide a value from the XMTR module's ADC.

What do these bytes tell you? First, the two digital inputs at the XMTR module are in the logic-1 state. Second, the ADC value, 0x01D7, translates to 471_{10} or 3.3 * (471/1024) volts, which comes to about 1.52 volts.

Step 7. Now you will force the XMTR module to transmit information. Remember you set the IC-DIO Change-Detect parameter to 0x08, which corresponds to the AD3-DIO3 input on the XMTR module. Changing the logic state of the AD3-DIO3 input should force the XMTR to transmit the digital-input logic states.

Please read the following lettered steps below before you perform them because they involve making an electrical change at a specific time.

• Clear the X-CTU Terminal window.
• Turn on power to the breadboard.
• After 10 seconds you should see a string of hex values appear in the Terminal window. As soon as this happens, count off four or five seconds.
• Then remove the XMTR connection between pin 17 (AD3-DIO3) and +3.3-volt power at the power bus and connect this end securely to ground. This change places a logic 0 on the AD3-DIO3 input at the XMTR module and should force a transmission of the logic states on the active logic inputs. This type of forced transmission will NOT include any analog-input information. Turn off power to the breadboard.

When ready, please perform the four steps above. You can repeat them several times. Just ensure you start with a logic 1 (+3.3 volts) applied at the AD3-DIO3 input (pin 17) and when you move this connection from +3.3 volts to ground ensure you have a good ground connection. Moving the jumper between +3.3 volts and ground provides a crude way to change logic states: logic 1 to logic 0. (The last section of this experiment, Make Clean Logic Transitions, explains a better way to create clean control signals.)

```
Block 1
7E 00 30 83 56 78 32 00 0A 02 0A 00
0A 01 D8 00 0A 01 D7 00 0A 01 D7 00
0A 01 D7 00 0A 01 D7 00 0A 01 D7 00
0A 01 D7 00 0A 0A D7 00 0A 01 D8 00
0A 01 D7 90

Block 2
            7E 00 18 83 56 78 32 00
04 02 0A 00 0A 01 D7 00 0A 01 D8 00
0A 01 D8 00 0A 01 D7 E2

Block 3
                  7E 00 0A 83
56 78 33 00 01 00 0A 00 02 6E

Block 4
                        7E 00
30 83 56 78 2D 00 0A 02 0A 00 02 01
D8 00 02 01 D8 00 02 01 D7 00 02 01
D8 00 02 01 D8 00 02 01 D7 00 02 01
D8 00 02 01 D8 00 02 01 D7 00 02 01
D8 E0
```

FIGURE 7.2 This data appeared in the Terminal window when I performed the sequence listed in Step 5. Dividing the data into blocks makes it easier to interpret.

The information in Figure 7.2 shows information from the X-CTU Terminal window after I performed the steps above. I separated this information into four blocks for clarity and underlined the analog and digital data sent from the XMTR module.

Each block starts with the standard start-of-transmission value (7E), number of bytes, module address, signal-strength byte, and option byte.

Block 1 shows the *first* transmission with 10 samples of digital and analog information (underlined).

Block 2 shows an abbreviated set of four (0x04) samples (underlined) taken prior to the change at the AD3-DIO3 input.

Block 3 shows only one sample, which identified only the active digital lines and indicated the logic level on the active digital lines (underlined). The communication ended with a checksum. Note that the Block-3 data indicates only one sample of digital inputs in the transmission.

Block 4 shows the next transmission of 10 samples of analog and digital information (underlined), gathered based on the configuration information saved in the XMTR module.

What happened at the XMTR module? After the first series of 10 samples appeared (Block 1), I waited from four to five seconds and then grounded the input to the AD3-DIO3 pin (pin 17). That transition from a logic 1 (3.3 volts) to a logic 0 (ground) caused the XMTR module to first transmit the samples it had acquired up until that logic transition. So we see the abbreviated set of four samples shown in Block 2.

Next, the XMTR module transmitted the forced sample taken in response to the logic transition at the AD3-DIO3 pin. Thus the information in Block 3 only identifies the active *digital* inputs and their state. Because I grounded the AD3-DIO3 input, it shows a logic 0 and the AD1-DIO1 input shows a logic 1. After the XMTR module sends the forced sample, it goes back to its programmed mode, accumulates 10 samples over 10 seconds, and transmits them.

Step 8. After you complete the steps above and analyze your data, you should have the AD3-DIO3 jumper connected to ground. If not, move it to ground now to place a logic 0 on the AD3_DIO3 input. Now perform the sequence described below. In this step, though, you will change the AD3-DIO3 input from a logic 0 to a logic 1.

- Clear the X-CTU Terminal window.
- Turn on power to the breadboard.
- After 10 seconds you should see a string of hex values appear in the Terminal window. As soon as this happens, count off four or five seconds.
- Then at the ground bus, remove the connection between ground and pin 17 (AD3-DIO3). This action changes the AD3-DIO3 input at the XMTR module to a logic 1. Because an unconnected XBee pin floats to a logic 1, you don't have to connect the jumper to +3.3 volts to produce a logic 1 at the AD3-DIO3 pin. Turn off power to the breadboard.

You should see an arrangement of information similar to that observed in Figure 7.2. This response indicates either a logic-1-to-logic-0 *or* a logic-0-to-logic-1 transition will force a transmission of data from active digital inputs at a transmitter.

Step 9. You could set up a transmitter to send, say, a single sample (IT – Samples before TX = 0x01) every 60,000 milliseconds (60 seconds) by setting IR-Sample Rate = 0xEA60. By using one digital input as a trigger and setting IC-DIO Change-Detect to let that input force a transmission, a switch or sensor could force a transmission at any time. Use this type of operation when you must receive digital-input information from a transmitter as soon as some event occurs.

If you plan to use this mode keep the following points in mind:

- You must have at least one digital *input* enabled with its Dx-DIOx Configuration set to 3-DI.
- You must set the corresponding DIOx bit to a logic 1 in the IC-Change-Detect byte programmed in the transmitting XBee module.
- The received information might include a "short" set of data that represents only samples already taken at the transmitter and ready for transmission. If you used a microcontroller to receive this data, its software must handle this type of abbreviated transmission.
- A forced transmission updates the receiver about the state of any active digital inputs. That means enabled digital-output pins at the receiver receive an update and the data from the receiver's UART also reflects the

current state of digital-input pins at the transmitter. In this experiment, the UART sends serial data to the X-CTU Terminal window, but you could send it to a microcontroller instead.

Step 10. Optional. The circuit shown in Figure 7.1 includes a jumper for the AD1-DIO1 input. Determine how to change the forced transmission from a logic transition on the AD3-DIO3 input to a change on the AD1-DIO1 input. Test your change. Did it work? Hint—you only need to change the IC – DIO Change Detect configuration.

Could you change the IC – DIO Change Detect configuration so a change on the AD1-DIO1 *or* the AD3-DIO3 input forces a transmission? How would you do it?

MAKE CLEAN LOGIC TRANSITIONS

You can change electronic logic states from a logic 1 (+3.3 volts) to a logic 0 (ground) by moving a wire between these two voltages. But this crude technique can cause problems when you need a single transition. Mechanical switches and wires to power or ground contacts tend to "bounce," or open and close rapidly, which causes several transitions between logic-1 and logic-0 conditions. The timing diagram in Figure 7.3 shows the transitions between a logic 1 and logic 0 caused by a mechanical switch within the circuit shown in Figure 7.4. You can identify many transitions from one switch closure.

When you need one clean transition per switch actuation, "debounce" a mechanical switch's contacts. You can use a cross-coupled NAND-gate integrated circuit such as an SN74AHC00 and wire it as shown in Figure 7.5. You will need two pull-up resistors and a single-pole double-throw (SPDT) switch or pushbutton.

An SN74AHC00 IC includes four separate NAND gates so you can debounce two switches with one IC. This IC can operate with a power-supply

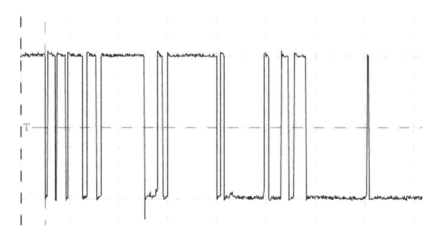

FIGURE 7.3 This oscilloscope display shows logic-1 to logic-0 transitions caused by contact bounce in a mechanical switch. The pulses have a 0-to-5-volt range over 1.6 msec.

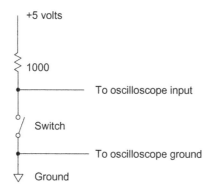

FIGURE 7.4 This simple circuit lets a storage oscilloscope capture a mechanical-switch-bounce signal.

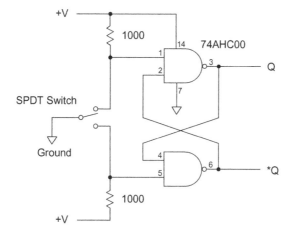

FIGURE 7.5 This debounce circuit can operate with a supply voltage between 2.0 and 5.0 volts, so you can use it with 5-volt or 3.3-volt logic circuits.

voltage between +2.0 and +5.5 volts, so by changing its supply voltage you can operate debounce circuits for either 3.3-volt or 5-volt logic circuits. With the switch in the position shown, the Q output remains at a logic 1 and the *Q output remains at a logic 0. When you actuate the switch, the Q output becomes logic 0 and the *Q output becomes a logic 1, without any extra transitions created by switch bounce. For more information about switch debouncing techniques, read, "A Guide to Debouncing," by Jack Ganssle at: http://www.ganssle.com/debouncing.htm.

How to Handle Data from Several Analog Inputs

REQUIREMENTS

2 XBee modules
2 XBee adapters
1 3.3-volt DC power supply
1 10-kohm (10,000-ohm) variable resistor
1 10-kohm resistor, 1/4-watt, 10% (brown-black-orange)
1 4700-ohm resistor, 1/4-watt, 10% (yellow-violet-red)
1 Solderless breadboard
1 USB-to-XBee adapter
1 USB cable—type-A to mini-B
Digi X-CTU software running on a Windows PC, with an open USB port
Small screwdriver, flat blade
Copies of the tables from Appendix I

INTRODUCTION

In previous experiments you learned how an XBee receiver formats its data. With that knowledge, you can separate, or parse, the received information into groups to determine the logic state of any active digital inputs and extract data for any active ADC inputs. So far, though, experiments have used only one analog input. In this short experiment, you will use two analog inputs and learn how a receiver formats their data. The same format applies to communications from a transmitter that has more than two active analog inputs.

Step 1. Before you change any wiring on the breadboard, turn off power to the breadboard circuits. The diagram in Figure 8.1 shows the needed connections. If you have any other components attached to the XMTR adapter, please remove them. Then make the connections shown in Figure 8.1.

63

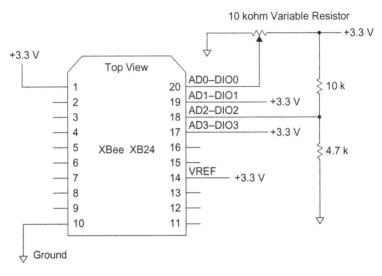

FIGURE 8.1 These connections provide two digital signals and two analog signals for the XMTR XBee module. Schematic diagrams now use a small triangle symbol to represent a common ground connection.

The two fixed resistors (10 kohms and 4700 ohms) create a voltage divider and supply a constant analog voltage of about 1.0 volts to the AD2-DIO2 (pin 18) input. Depending on the types of resistors you have, resistances can vary by about ±5 to ±10 percent, so the voltage on your XMTR module at pin 18 will likely come close to, but not equal to, 1.0 volts.

Place the XMTR module in the USB-to-XBee adapter and connect the adapter to the PC's USB cable. Run the X-CTU program and test communications with the XMTR module. In the Modem Configuration window, click on Read to obtain the configuration from the XMTR module. If you don't know the state of configuration settings, click on Restore and then click Read to obtain the default settings. In the X-CTU Modem Configuration window, set the XMTR-module configurations to:

DL – Destination Address Low = 1234
MY – 16-Bit Source Address = 5678
D0 – DIO0 Configuration = 2-ADC
D1 – DIO1 Configuration = 3-DI
D2 – DIO2 Configuration = 2-ADC
D3 – DIO3 Configuration = 3-DI
IT – Samples before TX = 03
IR – Sample Rate = 3E8

The Sample-before-TX value 0x03 will cause the XMTR to acquire three samples before it transmits them. New information will appear in the X-CTU Terminal window every three seconds.

Confirm these settings and save them in the XMTR module. You can find this configuration profile in EX8_XMTR_A.pro.

Step 2. Place the XMTR module in its adapter on the breadboard. Recheck the connections between the XMTR module and the breadboard.

Step 3. Insert the RCVR module in the USB-to-XBee adapter and reconnect this adapter to the USB cable. If you don't know the state of configuration settings, click on Restore and then click Read to obtain the default settings. Set the RCVR-module parameters to:

DL – Destination Address Low	= 5678
MY – 16-Bit Source Address	= 1234
D3 – DIO3 Configuration	= 5 D0-HIGH
P0 – PWM0 Configuration	= 2-PWM Output
IA – I/O Input Address	= FFFF

Confirm these settings and save them in the RCVR module. You can find this configuration profile in EX8_RCVR_A.pro. Leave the RCVR module in the USB-to-XBee adapter connected to the PC. You will use the X-CTU program to monitor data from the XMTR module.

Step 4. Set the variable resistor connected to the XMTR module to about mid range. Switch the X-CTU program to the Terminal window, clear the window, and set it to display hex values. Turn on power to the breadboard and watch as the X-CTU program displays data in hex characters. It should take three seconds for the first transmission to start. After you receive two or three transmissions, turn off power to the breadboard. You need only one or two sets of received information to analyze.

When I ran this experiment, I found the following hex values in the Terminal window. I reformatted the hex values to make them easier to understand. Your data will look different:

```
7E 00 1A 83 56 78 2C 00 03 0A 0A

00 0A 01 D8 01 48

00 0A 01 D8 01 48

00 0A 01 D8 01 48

E7
```

Here's how this hex information breaks down:

7E	= start of transmission
001A	= number of bytes in transmission
83	= code for 16-bit module addressing
5678	= 16-bit address
2C	= signal strength
00	= status byte
03	= number of samples
0A0A	= Active-Signal Bytes

Step 5. Use copies of the tables in Appendix I to decode the 0x0A0A value and determine which analog or digital inputs are active at the XMTR module.

You should find the AD0-DIO0 (pin 20) and AD2-DIO2 (pin 18) pins configured as active analog inputs and the AD3-DIO3 (pin 17) and AD1-DIO1 (pin 19) pins set as digital inputs. Because you have set up two ADC inputs at the XMTR module, the data displayed in the Terminal window will include a 2-byte sample for *each* ADC input. The digital and analog data conforms to the following format: digital-input data followed by ADC_x data, ADC_{x+1} data, and so on. The x subscript indicates the ADC data arrives first from the lowest-numbered ADC input, followed in order by data from the higher-numbered *active* ADC inputs. In this case you should see the ADC data for the AD0-DIO0 input arrive first, followed by the data from the AD2-DIO2 ADC. If a transmitter has *inactive* analog input pins (disabled, or used for digital I/O), results will not include any *analog* information for those pins.

Step 6. Again use the tables from Appendix I to decode one of the three samples of analog- and digital-input information in the six columns of hex values above or use the data you collected in the Terminal window. The first two bytes of my data, 0x00 and 0x0A, provide digital data, the next two bytes come from the first enabled ADC, and the final two bytes come from the second enabled ADC.

The AD3-DIO3 and the AD1-DIO1 inputs both connect to +3.3 volts, so, as expected, they both appear as a logic 1 in the digital-data information.

The four bytes of analog data, 0x01D8 and 0x0148, represent binary information from the XMTR module's internal 10-bit ADC. In this experiment, the 1.5 volts from the 10 kohm variable resistor connects to the AD0-DIO0 (pin 20) input. The fixed-resistor circuit provides about 1.0 volts to the AD2-DIO2 (pin 18) input. What voltages do the ADC bytes represent, based on a VREF input of +3.3 volts? Here are the results for my ADC values.

```
0x01D8 = 472₁₀ and (472/1024) * 3.3 volts = 1.52 volts at AD0-DIO0
input
0x0148 = 328₁₀ and (328/1024) * 3.3 volts = 1.06 volts at AD2-DIO2
input
```

These calculations confirm the XMTR transmitter sends the lowest-numbered ADC value first, followed by data from ADCs in numerical order. If in doubt, adjust the trimmer and observe the received values from the AD0-DIO0 input. Convert these values into voltages. Did you get the voltages you expected? The number of *pairs* of analog bytes transmitted always equals the number of ADC inputs you have enabled.

Step 7. In addition to changing the trimmer settings you also can swap the placement of the fixed 10-kohm and 4700-ohm resistors to change the voltage on the AD2-DIO2 (pin18) input and see how it affects the data.

Step 8. Here are some questions to answer: Which of the two analog inputs could control a PWM signal at a receiver? Which digital or analog inputs could force an immediate transmission? Could such a transmission include analog data? Why does the XMTR module need a +3.3-volt connection to the VREF input?

Find answers at the end of this experiment.

A REMINDER ABOUT SAMPLE RATES

The manual for the Digi International XBee/XBee-PRO RF modules notes the following:

> *The maximum sample rate you can achieve while using one ADC input is one sample every millisecond or 1000 samples/sec. Note that the XBee transmitter cannot keep up with transmission when you set IR and IT equal to 1. Configuring an XBee to sample at rates faster than once every 20 milliseconds is not recommended.*

In other words, suppose you need three active ADC inputs. In theory a transmitter needs at least three milliseconds to perform the conversions, because the analog inputs *share* the ADC. An XBee module will perform a conversion, store the result, and connect the ADC to the next active ADC input pin, and so on. The more ADC inputs you activate, the more time you must allow for sampling. I don't recommend you try a sample period shorter than the 20-millisecond period Digi International recommends. The sampling time does not increase as you enable more *digital* input or output pins.

ANSWERS TO QUESTIONS IN STEP 8

- Which of the two analog inputs could control a PWM signal at a receiver? Only the AD0-DIO0 and AD1-DIO1 analog inputs can control the PWM0 and PWM1 outputs, respectively.
- Which digital input or inputs could force an immediate transmission? A transition of a logic signal at any digital input can trigger an immediate transmission. You must configure the I/O pin needed for the trigger input as a digital input and you must place a 1 in the IC-Change-Detect configuration value for that digital input.
- Could such a transmission include analog data? A transmission forced by a transition at a digital input contains only information about digital inputs. It contains no information from analog inputs.
- Why does the XMTR module need a +3.3-volt connection to the VREF input? All analog-to-digital converters need a reference voltage that determines the range of the ADC. Without a stable reference, the ADC would produce meaningless results. The reference should offer better accuracy than the ADC it connects to.

Investigate Sleep Modes and Sleep-Mode Timing

REQUIREMENTS

2 XBee modules
2 XBee adapters
1 3.3-volt DC power supply
1 10-kohm (10,000-ohm) variable resistor
1 Solderless breadboard
1 USB-to-XBee adapter
1 USB cable—type-A to mini-B
Digi X-CTU software running on a Windows PC, with an open USB port
Small screwdriver, flat blade

INTRODUCTION

In this experiment you will learn how to place an XBee module in a "sleep" state that saves power. This type of operation comes in handy when you plan to operate a remote wireless device from battery power or from an energy-harvesting power source. The XBee modules include two sleep modes that reduce a module's power consumption, and you select the mode within the Modem Configuration window under the Sleep Modes [NonBeacon] heading:

SM – Sleep Mode   = 1-PIN HIBERNATE
SM – Sleep Mode   = 2-PIN DOZE

When in the DOZE mode used in this experiment, an XBee module will draw less than 50 microamperes (50 μA) and the module will wake up to a completely operational condition in two milliseconds (2 msec). When in the HIBERNATE mode, an XBee module will draw even less power—below 10 microamperes (10 μA), but it will take the module longer to awaken—13.2 milliseconds (13.2 msec). In this mode you decrease power another five fold,

69

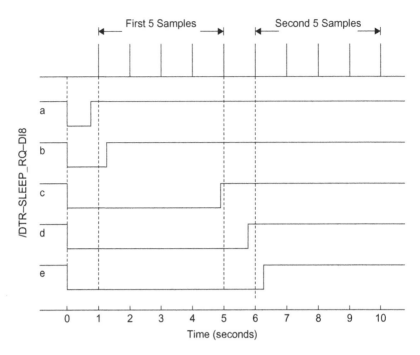

FIGURE 9.1 Timing relationships between a sleep-request signal at the /DTR-SLEEP_RQ-DI8 input and the information sampled by an XBee module. This diagram assumes a programmed sequence of five samples at one-second intervals.

but at the cost of a longer wake-up time. This experiment does not test the hibernate mode, but you can experiment with it on your own.

When in either of these two modes, a logic 1 on the /DTR-SLEEP_RQ-DI8 input keeps a module in a sleep condition. When the /DTR-SLEEP_RQ-DI8 input changes to a logic 0, the module wakes up and starts to sample information as previously configured with Sample Rate and Samples-before-TX values. A data transmission occurs as soon as the XBee module takes the last sample. The timing diagram in Figure 9.1 shows the relationship between five different logic-0 pulse widths at the /DTR-SLEEP_RQ-DI8 input and how they affect the operation of an XBee in a doze or hibernate mode. The descriptions that follow explain the effect of each pulse type.

When you look at the timing information in Figure 9.1 you see five logic-0 signals at the /DTR-SLEEP_RQ-DI8 input. Here's how they affect an XBee module in doze or hibernate mode based on my observations:

Pulse a. When the length of the logic-0 signal is shorter than the interval between samples, an XBee module does not sample any inputs and does not transmit information. This example uses a sample period of one second, so a pulse width of less than one second does not awaken a "sleeping" XBee module.

Pulse b. When the length of the logic-0 signal exceeds the interval between samples, an XBee module proceeds with the five samples and transmits the expected data.

Pulse c. When the length of the logic-0 signal exceeds the interval between samples, but does not extend to the final sample, an XBee module proceeds with the five samples and transmits the expected data.

Pulse d. When the length of the logic-0 signal exceeds the interval between samples, but does not extend to the time of the first sample in the successive set of samples, the XBee module transmits data for the first five samples. But it does not proceed with the second set of five samples and returns to its sleep state.

Pulse e. When the length of the logic-0 signal extends beyond the time of the first sample in the successive series of samples, the XBee module will complete the second set of samples and transmit their data.

In all five examples above, when an XBee module detects a logic 0 at its/DTR-SLEEP_RQ-DI8 input it comes out of the sleep mode and starts to draw a higher current. It will continue to draw this current until it transmits the last set of samples data and then goes back into a sleep mode. So, power consumption occurs as follows for the five types of pulses shown in Figure 9.1:

Pulse a. Full power used during the logic-0 period. No data transmitted.

Pulses b and c. Full power used during the five-sample period. Five samples transmitted.

Pulse d. Five samples transmitted and full power used during the *entire* logic-0 period.

Pulse e. Full power during the two five-sample periods. First five samples transmitted at the five-second mark, second five samples transmitted at the 10-second mark.

This timing and power-consumption information shows if you need a set of only five samples, keep the width of the logic-0 pulse applied to an XBee transmitter longer than the sample period and less than the sample period multiplied by the number of samples. So in this case: 1 sec </DTR-SLEEP_ RQ-DI8 < 5 sec. If you use a longer logic-0 pulse, you will get more data than you need and will increase power use by the transmitting XBee module.

Step 1. In this experiment you will set up the XMTR module so it has two digital inputs and one analog input. Before you change any connections, turn off power to the breadboard.

Make connections shown in Figure 9.2. If you have other components or wires that connect to the XMTR adapter, please remove them. To start, connect the /DTR-SLEEP_RQ-DI8 (pin 9) input to +3.3 volts to supply a logic 1 at pin 9. Later you will move this jumper wire to ground (logic 0).

Step 2. Place the XMTR module in the USB-to-XBee adapter and connect the adapter to the PC's USB cable. Run the X-CTU program and test communications with the XMTR module.

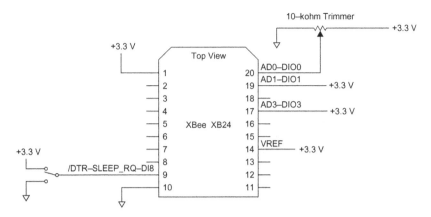

FIGURE 9.2 This XBee-module circuit includes a jumper that will control the power-saving sleep state.

In the Modem Configuration window, click on Read to obtain the configuration from the XMTR module. If you don't know the state of configuration settings, click on Restore and then click Read to obtain the default settings. In the X-CTU Modem Configuration window, set the XMTR-module configurations to:

DL – Destination Address Low = 1234
MY – 16-Bit Source Address = 5678
SM – Sleep Mode = 2-PIN DOZE
D0 – DIO0 Configuration = 2-ADC
D1 – DIO1 Configuration = 3-DI
D2 – DIO2 Configuration = 0-DISABLED
D3 – DIO3 Configuration = 3-DI
IT – Samples before TX = 0A
IR – Sample Rate = 3E8

Confirm these settings and save them in the XMTR module. You can find this configuration profile in EX9_XMTR_A.pro.

The Sample-before-TX value 0x0A will cause the XMTR to acquire 10_{10} samples before it transmits them. Thus new information will appear in the X-CTU Terminal window every 10 seconds.

Step 3. Place the XMTR module in its adapter on the breadboard and recheck the connections.

Step 4. Place the RCVR module in the USB-to-XBee adapter and reconnect this adapter to the USB cable. If you don't know the state of configuration settings, click on Restore and then click Read to obtain the default settings. Set the RCVR-module parameters to:

DL – Destination Address Low = 5678
MY – 16-Bit Source Address = 1234
IA – I/O Input Address = FFFF

Confirm these settings and save them in the RCVR module. You can find this configuration profile in EX9_RCVR_A.pro. Leave the RCVR module in the USB-to-XBee adapter connected to the PC. You will use the X-CTU program to monitor data from the XMTR module.

Step 5. Set the variable resistor connected to the XMTR module between 1/4 and 3/4 of the way between its end stops. Select the Terminal window, clear it, and set it to display hex values. Turn on power to the breadboard and wait. You should not see any received data in the Terminal window because the connection of the /DTR-SLEEP_RQ-DI8 (pin 9) input to logic 1 (+3.3 volts) keeps the XMTR module in the doze mode.

Step 6. Move the wire connected to the /DTR-SLEEP_RQ-DI8 (pin 9) input from logic 1 (+3.3 volts) to logic 0 (ground). The logic 0 on pin 9 will cause the XMTR module to exit the doze mode and transmit data *after* it collects all 10 samples. So, ten seconds after you change the /DTR-SLEEP_RQ-DI8 input to a logic 0, you should see data appear in the Terminal window.

As long as you keep the /DTR-SLEEP_RQ-DI8 input connected to a logic 0, the XMTR will continue to send 10 samples every 10 seconds. You could use a switch or a signal from another electronic device such as an MCU to change the logic state at the /DTR-SLEEP_RQ-DI8 input and awaken an XBee module.

Adjust the variable resistor and observe the data in the Terminal window. You should see the value of the bytes from the XMTR ADC change.

To stop the transmission of data, move the wire connected to the /DTR-SLEEP_RQ-DI8 input at pin 9 to logic 1. You might see one final burst of data in the Terminal window after you reconnect the /DTR-SLEEP_RQ-DI8 input to logic 1. You'll learn why in the next step.

Step 7. In this step you will see what happens when you try to switch the XMTR module into a sleep mode during the time it samples the two digital inputs and one analog input.

Suppose the XMTR module has sent data for 10 samples and it already has taken six samples to prepare for the next transmission. At this time you change the /DTR-SLEEP_RQ-DI8 input to logic 1 to force the XMTR module into its sleep mode. What happens to those six samples? Does the XMTR module save them for the next transmission? Does it continue sampling until it has all 10 samples, transmit them, and only then go into the sleep mode? Or does it discard them?

Use the variable resistor to change the voltage measured by the XMTR ADC so you can watch the effect of taking the XMTR out of its sleep mode and putting it back into sleep mode. Turn off power to your breadboard and clear the Terminal window. Change the /DTR-SLEEP_RQ-DI8 (pin 9) input to a logic 1 (+3.3 volts). This connection puts the XMTR in the sleep mode.

Do not perform the following lettered steps below yet. First read them so you understand what to do and the timing involved. Use two hands in this experiment.

- Turn on power to the breadboard. No information should appear in the Terminal window.
- Use a small screwdriver to slowly adjust the variable resistor back and forth between its end points. Continue to adjust the variable resistor back

and forth until instructed to stop. (You don't have to go all the way to the end points, though.)

• As you change the variable resistor setting, move the /DTR-SLEEP_ RQ-DI8 input at pin 9 to a logic 0. In 10 seconds, the first burst of information should appear in the Terminal window.

• As soon as you see the data appear in the Terminal window, count off four or five seconds and then change the /DTR-SLEEP_RQ-DI8 input back to a logic 1. This step signals the XMTR module to go back into its sleep mode after it has already taken four or five samples. (You don't need precise timing here; you just want to change the logic level at the /DTR-SLEEP_ RQ-DI8 input as the XBee module takes the second group of 10 samples.)

• You should see a second burst of data appear in your Terminal window.

Now stop adjusting the variable resistor. Turn off power to the breadboard. After you understand these steps, do them. Don't hesitate to try them several times. Clear the Terminal screen each time so you can easily see the newest data.

At the end of these steps, the Terminal window should show two sets of data—the data taken when the XMTR module awakened in step (c) above, and the data taken as you put the XMTR module back into its sleep mode in step (d).

My Terminal displayed the data shown next, which I reformatted for clarity. The two left-most columns represent digital information from the two digital inputs, DIO1 and DIO3. The value of these bytes (0x000A) did not change because I did not change the logic level on the corresponding pins. The two right-most columns represent the analog voltage (underlined), measured as I changed the variable-resistor setting.

First burst of data after wake-up:

```
7E 00 30 83 56 78 3D 00 0A 02 0A

00 0A 02 31

00 0A 00 00

00 0A 00 FB

00 0A 02 70

00 0A 03 FF

00 0A 03 08

00 0A 01 12

00 0A 00 00

00 0A 02 4D

00 0A 03 FF
```

E6

Second burst of data taken as I put the XMTR back into sleep mode but continued to vary the trimmer setting:

```
7E 00 30 83 56 78 2E 00 0A 02 0A

00 0A 03 04

00 0A 01 75

00 0A 00 00

00 0A 01 6D

00 0A 03 40

00 0A 03 A7

00 0A 01 CD

00 0A 00 0E

00 0A 00 2A

00 0A 02 06

20
```

Before you analyze this data, review the steps performed earlier: The XMTR module acquired the first 10 samples after you woke it up. Then, about four or five seconds into the next set of 10 samples, you signaled the XMTR module to go back into a sleep mode. In the second set of data, the analog voltage continued to vary even after you changed the /DTR-SLEEP_RQ-DI8 input to logic 1.

You expect to see the first set of data, but you might not have expected the XMTR module to send the second set of data. After all, you might think it would go into its sleep mode as soon as you changed the /DTR-SLEEP_RQ-DI8 input to a logic 1. But because the /DTR-SLEEP_RQ-DI8 input remained at a logic 0 beyond the minimum time needed to awaken—or keep awake—the XMTR module, it continued its task. So, you obtained a complete set of 10 samples. Refer back to Figure 9.1 for the timing relationships.

Step 8. In this step you will awaken the XMTR module and keep it awake only long enough to acquire one set of data. Turn off power to the breadboard.

Do not perform the following lettered steps yet. Read them first to understand what you will do and the timing involved. You need both hands in this experiment. The following steps *do not* simply duplicate those you performed earlier. Pay careful attention to the timing.

• Connect the /DTR-SLEEP_RQ-DI8 input at pin 9 to logic 1 (+3.3 volts). If the Terminal window contains information, click on Clear Screen to get a clear area. Turn on power to the breadboard. You should not see any information appear in the Terminal window.

- Use a small screwdriver to slowly adjust the variable resistor back and forth between its end points. Continue to adjust the variable resistor back and forth until instructed to stop. (You don't have to go all the way to the end points, though.)
- As you change the variable resistor setting, change the /DTR-SLEEP_RQ-DI8 input at pin 9 to a logic 0 to take the XMTR module out of the sleep mode. After about four or five seconds, change the /DTR-SLEEP_RQ-DI8 input back to a logic 1. This signals the XMTR module to go back into its sleep mode after it has sampled the two digital inputs and one analog input four or five times. (You don't need precise timing here, but you must put the XMTR module back in the sleep mode before the end of the 10-second sample period.)
- As soon as a burst of data appears in the Terminal window, stop adjusting the variable resistor.

Now you should see only one set of 10 samples.

To put an XBee module back in its sleep mode, you must return the /DTR-SLEEP_RQ-DI8 (pin 9) input to a logic 1 before the module completes its sequence of taking the number of samples you programmed as the Samples-before-TX parameter for the XMTR module. If you must cause a module to perform a short operation, using a switch to change logic levels on the /DTR-SLEEP_RQ-DI8 input might not work because you couldn't switch back to a logic-1 input quickly enough. You would have to use a short logic-0 pulse from a microcontroller or other device.

Step 9. Optional. You can experiment with an XBee module in the hibernate state by changing the SM-Sleep Mode setting to 1-PIN HIBERNATE, but you will not see any differences unless you can measure the XMTR module's current use and the module's start-up time when it comes out of a sleep mode.

Note: Rather than change the variable-resistor setting by hand, I also ran this experiment and used a triangle-wave generator (Global Specialties Model 2001) to provide a slowly changing voltage between 0 and 3 volts for the ADC. The results duplicated those shown earlier.

How to Use API Packets to Control a Local XBee Module

REQUIREMENTS

1 XBee module
1 USB-to-XBee adapter
1 USB cable—type-A to mini-B
Digi X-CTU software running on a Windows PC, with an open USB port

INTRODUCTION

In this experiment you will learn how to use application programming inter-face (API) packets that include AT commands to control an XBee module and transfer information to and from it. These commands, which use the stan-dard AT-command letters, give you greater control over modules than you can obtain with the Modem Configuration settings alone. And they let you break away from using the X-CTU software.

Although you connected an XBee module to a PC via a USB port, the X-CTU software handled communications through a virtual serial port. These communications use a device called a Universal Asynchronous Receiver/Transmitter (UART) that forms the core of a serial port. A UART can trans-mit and receive bytes of information at specific bit rates adopted by conven-tion, and it always formats data with a logic-0 *start bit* followed by n bits of data, where n usually equals eight, followed by a logic-1 *stop bit*, as shown in Figure 10.1. Although a UART can send a parity bit for error detection, most communications do not include it. Visit the Wikipedia Web site for an article that provides more details about UART operations: http://en.wikipedia.org/wiki/Universal_asynchronous_receiver/transmitter.

In a UART-type communication, each bit requires the same period. Thus, if you have a 9600 bits/second transmission, each bit takes 1/9600 seconds,

77

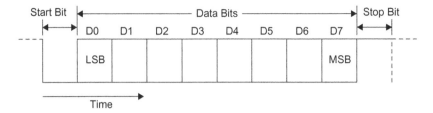

FIGURE 10.1 A UART receives and transmits 8-bit data in this standard format, with a logic-0 start bit and a logic-1 stop bit. Although some UART communications include a parity bit, most do not.

or about 104 microseconds. The receiver and transmitter must operate their UARTs with the same bit rate.

An integrated circuit in the USB-to-XBee adapter converts the USB signals into signals for the UART within the XBee chip. According to Digi International:

> *The API commands act much like the Modem Configuration settings, but your PC—or another device, such as a microcontroller (MCU)— can transmit API commands on the fly to both local and remote XBee modules. In the case of a microcontroller, communications would eliminate the USB connection and hardware would provide a direct UART-to-UART connection. Your microcontroller's code would create the API commands and send them to the XBee module, and the MCU would received responses triggered by API commands and interpret them.*

Almost every microcontroller (MCU) includes at least one UART, loosely called a serial port. The UART simplifies communications between an MCU and devices such as XBee modules. Later experiments will involve MCU-UART-to-XBee-module communications and control software.

After you finish this experiment, go on to the next one in the same session because it builds on the experience you gain here.

Step 1. In this step you will configure the RCVR module to operate with API packets sent from the X-CTU Terminal. Each packet includes information you have already learned about along with a 2-letter AT-modem-control command such as MY, SL, and IT. In some cases, the AT command includes data for the attached XBee module.

Place the RCVR module in the USB-to-XBee adapter and reconnect this adapter to the USB cable. Click on Restore and then click Read to obtain the default settings.

In the Modem Configuration window, look under the Networking & Security heading and note the hexadecimal serial numbers for your RCVR module:

SH – Serial Number High   = _____

SL – Serial Number Low   = _____

Digi has pre-programmed a unique serial number in each module and it will never change. Set the RCVR-module parameters to those shown next. Find these settings in the file: EX10_RCVR.pro:

DL – Destination Address Low = 5678
MY – 16-Bit Source Address   = 1234
IA – I/O Input Address   = FFFF

Find the Serial Interfacing heading and locate the API-Enable setting. Change its setting to:

AP – API Enable   = 1-API ENABLED

Confirm these settings and save them in the RCVR module. Leave the RCVR module in the USB-to-XBee adapter connected to the PC.

Step 2. Now you can send an AT command in an API packet to the RCVR module from the X-CTU program. First, though, you must understand how to create such a packet, which requires several bytes. (The Digi *XBee/XBee-PRO RF Modules* manual provides more information about the API packets that contain the AT commands and responses to AT commands.)

When you need to send an AT command to an XBee module connected directly to your computer, use a simple packet as shown below. All commands sent to a *local* XBee module use the same packet format. In this context, the word "local" refers to XBee modems connected directly to the packet sender, perhaps an MCU, or in this case, the X-CTU software:

0x7E   = start-of-transmission byte
0x____   = message-length bytes
AT-command structure goes here
0x__   = checksum

The basic AT-command structure of an API packet for a *local* XBee module includes:

- The identifier 0x08, which indicates a command for a local XBee module
- A frame-identifier value of your choice (all experiments use 0x52)
- The AT command
- Any information or data the AT command requires

Suppose you wanted to read the Serial Number Low from the attached RCVR XBee module. First create the AT command in the structure just introduced:

```
0x08 0x52 S L
```

The S and L characters come from the 2-letter settings listed in the X-CTU Modem Configuration window and also listed in Appendix C. Because you want to *read* the serial number, this command does not require any additional information. But you cannot just type letters such as S and L in an AT command. You must convert letters to their corresponding hex values: S = 0x53 and L = 0x4C.

Each character has a standard 8-bit binary value defined in the American Standard Code for Information Interchange, or ASCII. Tables available in Appendix G and on many Web sites let you find a character and its equivalent hexadecimal value in an ASCII table.

The final AT command looks like this: 0x08 0x52 0x53 0x4C. And you insert it into the API packet, which also assumes hex values:

```
7E ?? ?? 08 52 53 4C ??
```

The message-length value counts only the bytes in the message (underlined) and you only sum these bytes to calculate the checksum.

So, the command now looks like:

```
7E 00 04 08 52 53 4C ??
```

Next, calculate the *hexadecimal* checksum from: 0x08 + 0x52 + 0x53 + 0x4C = 0xF9. Then subtract this sum from 0xFF: 0xFF − 0xF9 = 0x06.

If you have a larger sum, such as 0xAFC7, just subtract the two least-significant digits, 0xC7, from 0xFF. An inexpensive calculator, such as the Casio FX-260 Solar or the Texas Instruments TI 36X Solar, can handle hex math. Appendix E includes information about how to use an Excel "Packet Creator" spreadsheet that lets you insert decimal and hexadecimal values and ASCII characters to create a hex packet complete with checksum. For more information about checksums and their limits, see Appendix B.

Now the complete API packet with the SL command looks like this, again in hex values:

```
7E 00 04 08 52 53 4C 06
```

Step 3. With the RCVR module connected to your PC, open the Terminal window in the X-CTU software. Click on Clear Screen. If you don't see a two-column window, click on Show Hex to see information as hexadecimal values.

Click on Assemble Packet. If you see information in the Send Packet window, click Clear. In the bottom-right corner of the Send Packet window, click HEX. Then, type in your API packet, but without spaces, which the Send Packet window includes automatically:

```
7E 00 04 08 52 53 4C 06
```

Check your data and then click on Send Data. You should see your API packet appear in the Terminal window in blue characters (7E...06) and the RCVR module's response in red characters (7E...F5). My computer displayed the information shown in Figure 10.2.

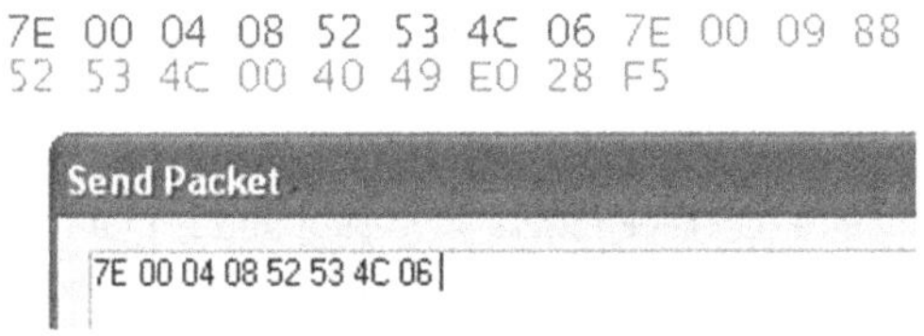

```
7E 00 04 08 52 53 4C 06 7E 00 09 88
52 53 4C 00 40 49 E0 28 F5
```

Send Packet

```
7E 00 04 08 52 53 4C 06
```

FIGURE 10.2 This screen image shows the packet sent to the RCVR module and the reply. The upper information appeared in the Terminal window.

If you did not see a response from your RCVR module, go back to Step 1 and check your settings. You must have the API mode enabled in the RCVR module. Use the X-CTU Modem Configuration section to read the current settings from the RCVR module so you can review them. Make any changes and then click Write to transfer the settings to the RCVR module.

Step 4. How do you interpret the reply from the RCVR module? It replied with information in a standard format. (Your data will vary from that shown next, but it will follow the same format.)

```
7E 00 09 88 52 53 4C 00 40 49 E0 28 F5
```

Here's how the hex information breaks down:

7E	= start of transmission
0009	= number of bytes in transmission
88	= packet type (AT Command Response)
52	= frame-identifier byte (always 52 in experiments)
53	= hex code for letter S
4C	= hex code for letter L
00	= status byte
4049E028	= SL information from my RCVR module
F5	= checksum for this message

In this example, the 0x88 identifies the message as a standard AT Command Response and the 0x52 duplicates the frame-identifier value sent to the RCVR module. This value—which you assign—should always match the frame-identifier byte sent in the original command packet. Although packets require this byte and it appears in responses, you do not have to do anything with it. The Status Byte identifies one of several conditions shown in Table 10.1. The response from the RCVR module indicates the OK condition.

Table 10.1 Status Bytes in XBee Communications

Status Byte	Status
0x00	OK
0x01	Error
0x02	Invalid Command
0x03	Invalid Parameter
0x04	No Response

Step 5. Again, click on Send Data to retransmit the API packet you created in Step 4 to ensure you get a response from the RCVR module.

Next, go to the packet in the Send Packet window and change the checksum value to: 0x00. This checksum is *not* valid for this packet but it lets you see how the RCVR module responds. The new packet should look like:

```
7E 00 04 08 52 53 4C 00
```

Click on Send Data. What did you observe? You should see no reply from the RCVR module. When a packet includes an incorrect checksum, the module that received the packet takes no action because the checksum it calculates differs from the checksum in your packet. Unfortunately, you can't tell what causes the lack of response. Always confirm the accuracy of a checksum before you transmit a packet. (If you plan to use an MCU to provide an API packet to an XBee module, your software can calculate the checksum, as later experiments demonstrate.)

Step 6. In this step you will send the RCVR module a packet with an invalid AT command, QQ, which does nothing. The letter Q corresponds to the hex value 0x51.

7E   = start byte
0004 = 4 bytes in message
08   = AT command-identifier byte
52   = frame-identifier byte
51   = hex code for Q
51   = hex code for Q
??   = checksum on four message bytes

Calculate the checksum you must use to replace the two question marks for the four bytes in the message.

```
7E 00 04 08 52 51 51 03
```

Clear the Send Packet window, type in the hex values shown directly above, and click Send Data. You should see the same data as shown below received from the XBee module:

```
7E 00 05 88 52 51 51 02 81
```

The information in Table 10.1 lets you interpret the Status Byte (0x02), which now indicates an Invalid Command.

Step 7. In this step, you will use the MY command to change the 16-bit Source Address in the RCVR module with the following API packet:

7E   = start byte
0006   = message length
08   = AT command-identifier byte
52   = frame-identifier byte
4D   = hex code for M
59   = hex code for Y
12AF   = hex values for Source Address
3E   = checksum

Clear the Send Packet window and type in the packet:

```
7E 00 06 08 52 4D 59 12 AF 3E
```

and click Send Data. You should see the reply below:

```
7E 00 05 88 52 4D 59 00 7F
```

You can parse this information as follows:

7E   = start byte
0005   = message length
88   = AT command response
52   = frame-identifier byte

4D   = hex code for M
59   = hex code for Y
00   = status byte
7F   = checksum

The Status Byte indicates no errors (see Table 10.1).

Switch to the X-CTU Modem Configuration window and click Read to obtain the configuration information from the RCVR. Under the Networking & Security heading, find the address given to MY – 16-Bit Source Address. You should see:

(12AF) MY – 16-Bit Source Address

An API packet that contains an AT command can change information within the RCVR module. You can use any of the AT commands in this way to modify or read the settings within an XBee module configured with the AP - API Enable set to 1-API ENABLED.

Be sure to change the 16-Bit Source Address (MY) back to 0x1234. Instead of using the Modem Configuration window, create an API packet to send the RCVR module via the Send Packet window.

Step 8. Optional. What would happen when you try to set the 16-bit Source Address but you use the hex codes for the lowercase letters m and y? Try this on your own.

0x6D   = Hex code for m
0x79   = Hex code for y

See my results at the end of this experiment.

Note: In a real-world design, most likely a microcontroller would issue AT commands within an API packet to make changes or read values. This type of MCU-to-XBee communication uses the same UART input and output pins used by the USB-to-XBee adapter. The adapter board includes an integrated circuit that handles the UART-to-USB communications.

In the next experiment you will learn how to use the API to change settings in a remote XBee module.

ANSWERS TO QUESTIONS IN STEP 8

The following API packet uses the lowercase "m" and "y" to change the MY address in an XBee module to 0x0000:

```
7E 00 06 08 52 6D 79 00 00 BF
```

When I transmitted this API packet, the RCVR module responded just as it would for the upper-case letters. A look at the binary code for the letters shows the relationship between the codes for upper- and lower-case letters:

```
M    = 0100 1101    0x4D

m    = 0110 1101    0x6D

Y    = 0101 1001    0x59

y    = 0111 1001    0x79
```

The only difference between the upper- and lower-case binary codes exists at bit position D5 underlined above (remember, start numbering bits with D0 for the right-most bit).

So, either an XBee module or the X-CTU software knows the codes for upper- and lower-case letters, or it simply ignores the D4 bit in ONLY the AT commands. The D4 bit is still valid in all other values in an API packet. I recommend you always use the ASCII values for uppercase letters in AT commands.

How to Use API Packets to Control Remote XBee Modules

REQUIREMENTS

2 XBee modules
1 XBee adapter
1 Solderless breadboard
1 USB-to-XBee adapter
1 USB cable—type-A to mini-B
Digi X-CTU software running on a Windows PC, with an open USB port

INTRODUCTION

In this experiment you will learn how to use application programming interface (API) packets to transfer information to and from local and remote XBee modules. The API lets you send AT commands to modules to control them via a wireless connection. If you have not performed Experiment 10, I recommend you do so before you proceed with this experiment. This experiment will take time to do properly, so don't rush and double-check your actions.

In previous experiments, you used the X-CTU software to configure an RCVR and an XMTR module so inputs at the XMTR could control outputs on the RCVR module connected through a USB cable to a PC. Now you will use application programming interface (API) packets to transmit AT commands from the RCVR module to the XMTR module. Although you labeled one module as a transmitter (XMTR) and one as a receiver (RCVR), both can act as transceivers and transmit and receive information, but not simultaneously. You will continue to use the XMTR and RCVR labels for clarity and consistency, even though the XMTR module might operate as a receiver and the RCVR module might act as a transmitter.

The Hands-on XBee Lab Manual.

Step 1. In this step you will configure the XMTR module to operate with the AT commands via the API. Ensure you have turned off power to the breadboard. Then place the XMTR module in the USB-to-XBee adapter and reconnect this adapter to the USB cable and your PC.

Within the X-CTU window, click on Modem Configuration and then click on Restore. After a few seconds you should see the message "Restore Defaults..complete" appear at the bottom of the X-CTU window. Click on Read to obtain the restored default values from the XMTR module.

Step 2. Check to ensure the DL – Destination Address Low and MY – 16-Bit Source Address each have a value of 0. This condition indicates the X-CTU software has reset all the XMTR module settings to their default state. If DL and MY are not equal to 0, perform Step 1 again.

Step 3. In the Modem Configuration window, look under the Networking & Security heading and note the serial number for your XMTR module. Digi has pre-programmed a unique serial number in each module and it will never change:

SH – Serial Number High   = _____

SL – Serial Number Low   = _____

Also note the value for SC – Scan Channels = _____

Step 4. Find the XMTR configuration in EX11_XMTR.pro. In the Modem Configuration window and under the Networking and Security heading, set the following hex values for the XMTR module:

DL – Destination Address Low   = 1234

MY – 16-Bit Source Address   = 5678

And under the Serial Interfacing heading locate the AP – API Enable setting. Change this setting to:

AP – API Enable = 1-API ENABLED.

Under I/O Settings heading look for the I/O Line Passing file icon and click on the + to view the settings. Click on the IA – I/O Input Address name and then click on the Set button that appears to its right. You should see a Set Hex String window open. In this window, click on Clear, type FFFF in the text area, and click on OK. In the Modem Configuration window you should see:

[FFFF] IA – I/O Input Address

Step 5. Recheck the four settings you changed in the Modem Configuration window. After you confirm the proper values, click on Write to save these settings in the XMTR module. After you see the message "Write Parameters...Complete" appear at the bottom of the X-CTU window, continue to the next step.

Step 6. Remove the USB-to-XBee adapter from the USB cable and remove the XMTR module from the adapter socket. Place the XMTR module in its XBee adapter on the breadboard.

Step 7. Place the RCVR module in the USB-to-XBee adapter and reconnect this adapter to the USB cable. Within the X-CTU window, click on Modem Configuration and then click on the Restore button. After a few seconds you should see the message "Restore Defaults..complete" appear at the bottom of the X-CTU window. Click on Read to obtain the restored default values from the RCVR module.

Step 8. Check to ensure the DL – Destination Address Low and MY – 16-Bit Source Address each have a value of 0. This condition indicates the X-CTU software has reset all the RCVR module settings to their default state. If DL and MY are not equal to 0, perform Step 7 again.

Step 9. Find the RCVR configuration in EX11_RCVR.pro. In the Modem Configuration window and under the Networking & Security heading, set the following values for the RCVR module:

DL – Destination Address Low = 5678

MY– 16-Bit Source Address   = 1234

And under the Serial Interfacing heading locate the AP – API Enable setting. Change this setting to:

AP – API Enable = 1-API ENABLED

Under I/O Settings heading look for the I/O Line Passing file icon and click on the + to view the settings. Click on the IA – I/O Input Address name and then click on the Set button that appears to its right. You should see a Set Hex String window open. In this window, click on Clear, type FFFF in the text area, and click on OK. In the Modem Configuration window you should see:

[FFFF] IA – I/O Input Address

Step 10. Recheck the four settings you changed in the Modem Configuration window. After you confirm the proper values, click on Write to save these settings in the RCVR module. Leave the RCVR module in the USB-to-XBee adapter socket connected to the PC.

Step 11. The schematic diagram in Figure 11.1 shows the only two connections—+3.3-volt power and ground—needed for the XMTR module in your breadboard. If you have other connections for the XMTR module, please disconnect them now.

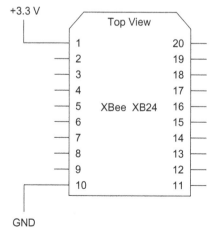

FIGURE 11.1 In this experiment, the XMTR module in the solderless breadboard needs only power and ground connections.

Step 12. Now you will learn how to send an AT command to the remote XMTR module. (The Digi International document, XBee/XBee-PRO RF Modules, provides more information about the AT commands, command packets, and the use of the API.)

When you need to send an AT command to a remote XBee module via a wireless link, you use the API-packet framework shown below. This format duplicates the one used in Experiment 10. The AT command structure forms the "payload" of the API packet.

0x7E   = start byte
0x____   = message length
Insert AT-command-specific structure here
0x__   = checksum

Depending on the AT command, the structure can vary. Information in a *remote* AT command includes:

*   An identifier of 0x17, which indicates a *remote* AT-command request
*   A frame-identifier value of your choice (all experiments use 0x52)
*   The 64-bit destination address; that is, the remote module's serial number
*   The destination module's 16-bit network address (MY), or 0xFFFE
*   A Command Options byte
*   The AT command
*   Any parameter the AT command requires

Important: In the section immediately above, the *remote* AT Command request uses an identifier byte of 0x17. In Experiment 10, the AT Command for a *local, directly connected module* used an identifier byte of 0x08.

My XMTR module has the serial number: 13A200 4049E0EC. So I used this address to identify it as the recipient for an AT command. Likely your Serial Number High value for the transmitter module also will appear as a 3-byte value because the X-CTU program does not display leading zeros. In that case, just append 0s to the left side of the Serial Number High value to create a 4-byte value. My Serial Number High appeared as (13A200) in the Modem Configuration window, so I changed it to: 0013A200.

Next, I created a remote-AT-command request to have my XMTR module in the breadboard send its Serial Number Low data back to the RCVR. Of course I already know the Serial Number Low information on my XMTR module, as would you. So this section of the experiment simply demonstrates how to create and use an API packet for a remote XBee module. The SL command returns already-known information, which makes it easy to confirm the remote command worked.

Here's how the remote-AT command (underlined information) would appear when placed in an API packet:

7E   = start byte
000F   = number of bytes in transmission
<u>17</u>   = Remote AT Command Request byte
<u>52</u>   = frame identifier value (all experiments use this
                      value)

<u>0013A2004049E0EC</u>   = XMTR serial number (SH and SL)
<u>FFFE</u>   = FFFE for 64-bit addressing with SH and SL bytes
<u>02</u>   = immediate action (explained later)
<u>53</u>   = hex code for letter S
<u>4C</u>   = hex code for letter L
EE   = checksum

In the Terminal window, I cleared the Send Packet window and typed in the packet below.

```
7E 00 0F 17 52 00 13 A2 00 40 49 E0 EC FF FE 02 53 4C EE
```

After I turned on power to my XMTR module, I clicked on Send Data and saw the response shown in Figure 11.2.

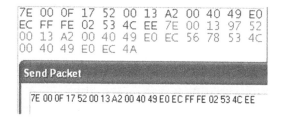

FIGURE 11.2 This figure shows the API packet sent to the XMTR module as well as the information received from the XMTR module in response to the AT command SL.

Step 13. XBee modules reply with information in a standard format. (Soon when you run this experiment for yourself, your data will vary from that shown because your XMTR module has a different serial number.)
Message received:

```
7E 00 13 97 52 00 13 A2 00 40 49 E0 EC 56 78 53 4C 00 40 49 E0
EC 4A
```

Here's how the hex information breaks down:
7E   = start byte
0013   = number of bytes in message
97   = packet type (remote Command Response)
52   = frame-identifier byte (all experiments use this value)
0013A200   = SH for responding module
4049E0EC   = SL for responding module
5678   = 16-bit address of responding module
53   = hex code for letter S
4C   = hex code for letter L
00   = status byte
4049E0EC   = SL value response to command
4A   = checksum

In this example, the reply included information in a standard format similar to that you have seen before. After the status byte, the next four bytes provide the Serial Number Low (SL) value for my RCVR module 0x4049E0EC, which corresponds to the Serial Number Low information in the X-CTU Modem Configuration window when I read the configuration information from my XMTR module and the SL information used in the API packet. So, the command worked at the remote XMTR module.

A quick review: I created a remote AT Command request sent to the RCVR module from the X-CTU program. The XMTR module picked up the wireless message from the RCVR and because its 64-bit serial number matched the serial number in the command, it executed the command and replied with its SL information.

Step 14. Now use the RCVR module and the X-CTU software to send the SL command to your XMTR module. Substitute your 8-byte XMTR module serial number (SH and SL) in the AT command framework below and calculate a new checksum marked "??" for the underlined values:

```
7E 00 0F 17 52 [SH] [SL] FF FE 02 53 4C ??
```

Remember to take the two least-significant hex digits from the sum and subtract only them from FF to yield the checksum. If your hex sum comes to 0xCA6, for example, subtract only 0xA6 from 0xFF, as in: $0xFF - 0xA6 = 0x59$.

Note: Do not include spaces—or the hex value for a space (0x20)—in transmissions you send to a module. Spaces help us visualize bytes, but a transmission simply sends one byte after another without any spaces.

Step 15. In the X-CTU window, click on Terminal and click Clear Screen. Ensure you have this window set for the 2-column display of hex values. Click on Assemble Packet to open the Send Packet text window. Click on Clear in the bottom-right corner of the Send Packet to ensure you have a clean text area. Also click on HEX in the same area to ensure the text you type appears as hex characters.

Move the cursor into the Send Packet text window and type your line of hex characters shown above but with *your substituted XMTR module's serial number and the checksum you calculated.* Recheck your typed values and correct any errors.

Step 16. Turn on power to your XMTR module, wait a few seconds, and in the Send Packet window, click on Send Data. You should see your transmitted packet in blue type and the response from your XMTR module in red type. If you do not get a response, or if you get something other than the response you expected, check the following troubleshooting tips:

• Did you use the correct serial-number information for *your* XMTR module? Remember you must have eight bytes of serial-number information. If your Serial Number High appears shorter than four bytes (eight hex digits), just place 0s in the most-significant positions—on the left end of the Serial Number High—to create a four-byte (8 hex-digit) value. If for example your Serial Number High (SH) appears as [095AC3] in the Modem Configuration window, place zeros on the left end to create eight hex digits in all: 00095AC3.

- Recheck your typing.
- Does the number of bytes in your message match the byte-count at the start of the message—the two-byte value that follows the 0x7E Start Byte.
- Did you calculate the proper checksum? Remember, don't include the Start Byte or the number-of-bytes hex values in your sum.
- Have you set the proper configuration for the XMTR and RCVR modules?
- Have you applied power to the XMTR module?
- Is the RCVR module attached to the USB-to-XBee adapter board and is that board connected to your PC?

Did you see the Serial Number Low (SL) in the information received by your RCVR module and displayed in the X-CTU Terminal window? Use the information in Step 12 to parse the response into sections you can interpret.

Step 17. In this step you will learn what happens when an XBee module transmits a command and the addressed module does not respond. Turn off power to the breadboard so the XMTR module cannot operate. Wait for a few seconds to let the power-supply voltage drop to 0. Go to the X-CTU Terminal window and clear it. If you don't see the Send Packet window, click on the Assemble Packet button.

Click on the Send Data button to transmit the message you created in Step 13 to the RCVR module. What do you see? I saw the information presented in Figure 11.3.

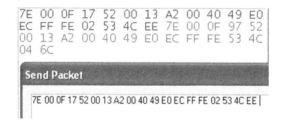

FIGURE 11.3 Sending the packet shown above to an unpowered XBee module results in a short message that indicates an error condition; the 0x04 byte just before the checksum.

In this case—for the unpowered XMTR module—the message returned came from the RCVR module with the status byte set to 0x04, which indicates No Response (see Step 12 or Table 10.1). The RCVR module transmitted the packet containing the SL command to my unpowered XMTR module, but it could not respond. When you connect an MCU to an XBee module, your software can examine the value of the Status Byte and take an action, such as sending you an alert message or trying another transmission.

Step 18. Now you will create a remote-AT-command request that addresses your XMTR module to have it return its *SC – Scan Channels value*, which you wrote down earlier in Step 3. But instead of using a long 64-bit address (SH and SL), you will use the 16-bit source address (MY) you established in the Modem Configuration window under the Networking & Security heading. Here's how you do it:

In the following packet, I replaced the serial number for the XMTR module with eight all-zero bytes, 0x00. Instead of placing 0xFFFE in the position left for the MY address information, I inserted the actual MY bytes for the XMTR module: MY = 0x5678. Then I used that address to create a remote-AT-command request as shown below:

7E	= start byte
000F	= number of bytes in transmission
17	= Remote-AT-command-request byte
52	= frame identifier value (all experiments use this value)
0000000000000000	= No serial number used, insert eight 0x00 bytes instead
5678	= 5678 = source address for XMTR module
02	= immediate actions (explained later)
53	= hex code for letter S
43	= hex code for letter C
30	= checksum

When you choose to identify a remote module with its 16-bit MY address, always insert eight 0x00 bytes for the serial number and insert the 2-byte MY address below it. The source address takes the place of the 0xFFFE information that occupied these two bytes when you used the remote module's 64-bit (8-byte) address in the command sequence shown in Step 12.

You can use 64-bit addressing at any time by simply placing the 64-bit (8-byte) address in the packet and following it with 0xFFFE in place of the MY address. The 0xFFFE value indicates to XBee modules that 64-bit addressing is in effect.

In this example, the AT command SC should return the Scan Channels value from the XMTR module.

Your and my XMTR modules have the same source address, 0x5678, so we can send the same packet to our XMTR module. Even though Digi has assigned each module a unique serial number, you can create your own unique 16-bit identifier and use it instead to address an XBee module. The complete API packet follows:

```
7E 00 0F 17 52 00 00 00 00 00 00 00 00 56 78 02 53 43 30
```

Clear the Terminal window and ensure you have power applied to the remote XMTR module. Set the Terminal window to display hex values and click on the Assemble Packet button. Clear the Send Packet window and type the hex values shown above. After you have checked your values and corrected any errors, click Send Data.

On my computer, I saw the information shown in Figure 11.4.

You will see standard information in response to the API packet followed by:

5678	= 16-bit MY address of XMTR module
53	= hex code for letter S

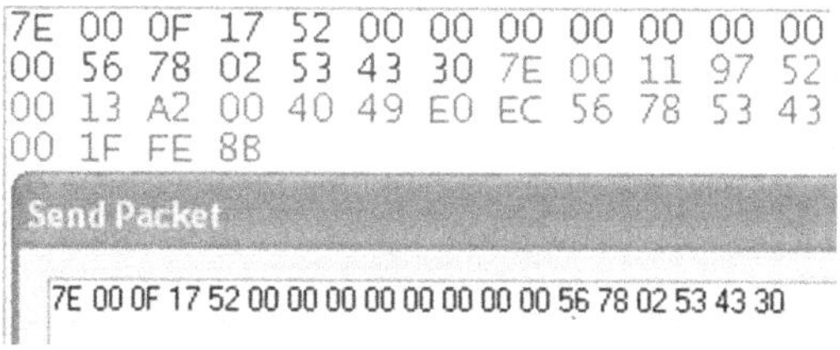

```
7E 00 0F 17 52 00 00 00 00 00 00 00
00 56 78 02 53 43 30 7E 00 11 97 52
00 13 A2 00 40 49 E0 EC 56 78 53 43
00 1F FE 8B
```

Send Packet

7E 00 0F 17 52 00 00 00 00 00 00 00 00 56 78 02 53 43 30

FIGURE 11.4 This screen image shows the API packet with the SC AT command embedded in it and the response from the XMTR module.

43 = hex code for letter C
00 = status byte (status = OK)
1FFE = requested Scan-Channel data from responding XMTR module

Step 19. Could you create an API packet to obtain the SC information by using the complete address—SH and SL information—for your XMTR module instead of the 16-bit address? Remember to set the 16-bit MY address in the packet to 0xFFFE. Try it.

Step 20. If you plan to go on to Experiment 12, you can leave your XMTR module connected to the solderless breadboard and the RCVR module attached to the USB-to-XBee adapter that connects to your PC. That experiment shows how to use AT commands in API packets to control outputs at a remote XBee module.

IMMEDIATE COMMAND ACTIONS

In this experiment, each AT command sent to a remote module included:
0x02 = immediate actions

This Command Options value forces a remote module to immediately perform the action requested by the accompanying AT command. If you substitute 0x00 for this value, the receiving module postpones action until it receives an Apply Changes (AC) command in a separate API packet. I recommend you include the 0x02 Command Options value to cause changes to take effect immediately. Then you don't have to remember to send a separate AC command later.

How to Use API Packets to Control Remote I/O Lines

REQUIREMENTS

2 XBee modules
1 XBee adapter
1 Solderless breadboard
2 LEDs
2 220-ohms resistors, 1/4 watt, 10% (red-red-brown)
1 USB-to-XBee adapter
Digi X-CTU software running on a Windows PC, with an open USB port

INTRODUCTION

In this experiment you will learn how to use AT commands in application programming interface (API) packets to control the output lines on a remote XBee module. If you have not performed Experiments 10 and 11 I recommend you do so before you proceed with this experiment.

In Experiments 10 and 11 you used the API packets and AT commands to read values from and write values to both a local and a remote XBee module. The local module (RCVR) connected to a nearby PC, but the remote module (XMTR) used the wireless link to receive commands and transfer information. Now you will learn how to use AT commands to control output pins on a remote module.

By using the AT commands a microcontroller (MCU) can directly control remote devices based on instructions in a program or results from math or logic operations in software. If, for example, a remote XBee module provides temperature information, another XBee module could receive the temperature information and an attached MCU could decide whether to turn on a fan

95

or a heater, or generate an alarm signal. Then the MCU could transmit an AT command to another remote XBee module to control the fan or heater, or possibly shut off both. The use of a wireless command to control remote devices gives equipment designers great flexibility when they face a remote-control problem.

If you just completed Experiment 11 and have your modules still in place, please go to Step 3.

Step 1. In this step you will configure the XMTR module to operate with the AT commands via the API. Ensure you have turned off power to the breadboard. Then place the XMTR module in the USB-to-XBee adapter and reconnect this adapter to the USB cable and your PC. Find the XMTR configuration in EX12_XMTR.pro.

Within the X-CTU window click on Modem Configuration and then click on Restore. After a few seconds you should see the message "Restore Defaults..complete" appear in the message area at the bottom of the X-CTU window. Click on Read to obtain the restored default values from the XMTR module.

Check to ensure the DL – Destination Address Low and MY – 16-Bit Source Address each have a value of 0. This condition indicates the X-CTU software reset all the XMTR module settings to their default state. If DL and MY are not equal to 0, perform Step 1 again.

In the Modem Configuration window, set the following values for the XMTR module:

DL – Destination Address Low = 1234
MY – 16-Bit Source Address   = 5678
AP – API Enable   = 1-API ENABLED
IA – I/O Input Address   = FFFF

Save these configuration settings in the XMTR module. After you see the message "Write Parameters..Complete" in the bottom text window, remove the USB-to-XBee adapter from the USB cable and remove the XMTR module from the adapter socket. Replace the XMTR module in its XBee adapter socket on the breadboard.

Step 2. Place the RCVR module in the USB-to-XBee adapter and reconnect this adapter to the USB cable. Find the RCVR configuration in EX12_RCVR.pro.

Within the X-CTU window, click on Modem Configuration and then click on Restore. After a few seconds you should see the message "Restore Defaults..complete" appear in the message area at the bottom of the X-CTU window. Click on Read to obtain the restored default values from the RCVR module.

Check to ensure the DL – Destination Address Low and MY – 16-Bit Source Address each have a value of 0. This condition indicates the X-CTU software reset all the RCVR module settings to their default state. If DL and MY are not equal to 0, perform Step 2 again.

In the Modem Configuration window, set the following values for the RCVR module:

DL – Destination Address Low = 5678
MY – 16-Bit Source Address   = 1234
AP – API Enable   = 1-API ENABLED
IA – I/O Input Address   = FFFF

Save these configuration settings in the RCVR module. Leave the RCVR module in the USB-to-XBee adapter socket connected to the PC.

Step 3. The schematic diagram in Figure 12.1 shows the connections needed for the XMTR module in your breadboard. If you have connections other than 3.3-volt power (pin 1) and ground (pin 10) on the XMTR module, disconnect them now. Connect the LEDs and resistors to the AD2-DIO2 pin (pin 18) and the AD3-DIO3 pin (pin 17).

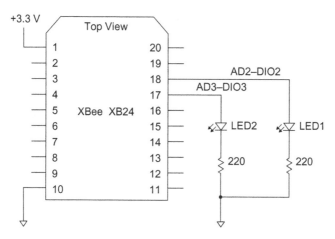

FIGURE 12.1 In this experiment you must connect an LED and a resistor in series for two I/O pins.

Step 4. Now you will learn how to send a command packet to the remote XMTR module to turn the LEDs on or off. To briefly review, an API packet uses the basic framework shown below.

0x7E = start byte
0x__ = message length
Insert AT-command-specific structure here
0x__ = checksum

Depending upon the AT command you choose to send, the AT-command-specific structure varies slightly. Use a *local* AT command to control an XBee module attached to your PC or an MCU. Use a *remote* AT command to send the command to a remote module you specifically address. Information in a *remote* AT command includes:

• An identifier of 0x17, which indicates a *remote* AT-command request
• A frame-identifier value of your choice (all experiments use 0x52)

- The 64-bit destination address; that is, the remote module's serial number
- The destination module's 16-bit network address (MY)
- A Command Options byte
- The AT command
- Any parameter the AT command requires

In Experiment 11 you learned how to use a remote module's 16-bit MY address to select it, rather than use its longer 64-bit serial number. This experiment continues the MY type of addressing. Your and my XMTR module have the same MY address of 0x5678, which we set in the Modem Configuration window and programmed into our XMTR module.

To control the AD2-DIO2 (pin 18) line for LED1, you use the the D2-DIO2 Configuration command and follow it with the condition you want to set for that pin. You can have one of five settings for I/O pins AD0-DIO0 through AD5-DIO5, as shown in Table 12.1. The AD6-DIO6 and AD7-DIO7 pins have digital I/O functions, too, but they lack the ADC capability and have other functions instead.

Table 12.1 I/O Pin Configurations for Remote AT Commands

I/O Parameter	Configuration
0x00	Disabled
0x01	Do Not Use
0x02	ADC - Analog Input
0x03	DI - Digital Input
0x04	DO - Digital Output (Low)
0x05	DO - Digital Output (High)

A digital-output-low condition forces the corresponding pin to ground, or logic 0. A digital-output-high condition forces the corresponding pin to approximately 3 volts, or logic 1. To set the D2 pin as a digital output in a logic-1 state (LED on), you use the D2 command and follow it with the value 5 as shown in the API packet that follows.

Important: To transmit a D2 command, for example, you use the ASCII hex value for the letter D (0x44) and the ASCII hex value for the numeral 2 (0x32). Do not enter the letter D or the numeral 2. They will make no sense to an XBee module in API mode.

7E	= start byte
0010	= number of bytes in transmission
17	= Remote AT Command Request byte
52	= frame identifier value
0000000000000000	= XMTR serial number (set to 0s)
5678	= 5678 = source address for XMTR module

02 = Command Options byte (immediate)

44 = hex code for letter D

32 = hex code for numeral 2

05 = hex code for I/O pin action (see Table 12.1)

4B = checksum

Step 5. Turn on power to your XMTR module and the LED circuits. In the Terminal window, clear the Send Packet window and type in the command packet below:

7E 00 10 17 52 00 00 00 00 00 00 00 00 56 78 02 44 32 05 4B

Transmit it to your powered XMTR module. What did you observe? LED1 should turn on and you should see a response in the Terminal window similar to that shown in Figure 12.2.

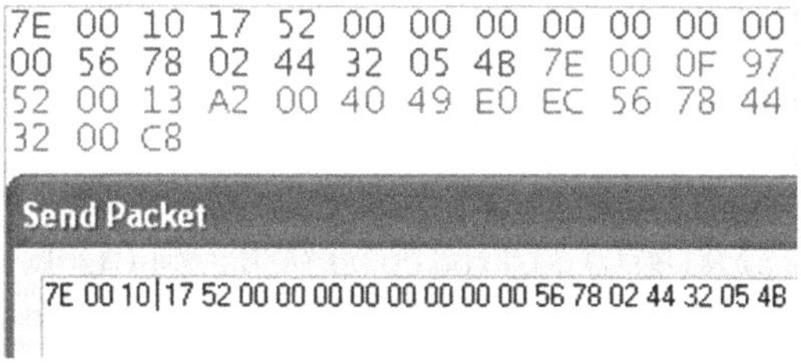

```
7E 00 10 17 52 00 00 00 00 00 00 00
00 56 78 02 44 32 05 4B 7E 00 0F 97
52 00 13 A2 00 40 49 E0 EC 56 78 44
32 00 C8
```

Send Packet

7E 00 10|17 52 00 00 00 00 00 00 00 00 56 78 02 44 32 05 4B

FIGURE 12.2 An API packet to turn on an LED at the AD2-DIO2 output pin caused the XMTR to transmit the response shown here.

Here's how the hex information breaks down for the latter part of the reply message:

7E 00 0F 97 52 00 13 A2 00 40 49 E0 EC 56 78 44 32 00 C8

...

5678 = 16-bit address of XMTR module

44 = hex code for letter D

32 = hex code for numeral 2

00 = status byte (OK)

C8 = checksum

To review: The response includes the D2 command you sent followed by the status byte, which in this case indicated OK.

Step 6. Can you create an API packet that includes an AT command to turn off LED1 at the AD2-DIO2 pin?

Here is a basic packet framework for you:

7E 00 10 17 52 00 00 00 00 00 00 00 00 56 78 02 ___ ___ ___ ___

See the answer at the end of this experiment. In the framework above, insert the hex bytes for the command and for the I/O action you want to occur. Then calculate the checksum for the 16 AT-command bytes. Remember to take the two least-significant hex digits from the sum and subtract only them from FF to yield the checksum. (You can turn off LED1 in more than one way, as explained later.)

Step 7. Send your packet to the XMTR module. You should see your transmitted packet in blue type and the response from your XMTR module in red type in the X-CTU Terminal window. Your LED1 should turn off.

If you do not get a response, or if you get something other than the response you expected, check the following troubleshooting tips:

- Recheck your typing.
- Does the number of bytes in your message match the byte-count at the start of the message—the two-byte value that follows the 0x7E start byte.
- Did you calculate the proper checksum? Remember, don't include the start byte or the number-of-bytes values in your sum. On some hexadecimal calculators, it's easy to confuse a lower-case b with the numeral 6.
- Have you set the proper configuration for the XMTR and RCVR modules?
- Have you applied power to the XMTR module? Have you properly connected the LEDs and resistors?
- Is the RCVR module attached to the USB-to-XBee adapter board and is that board connected to your PC?

Step 8. To turn off LED1, I used the D2 04 command and the following packet:

7E 00 10 17 52 00 00 00 00 00 00 00 00 56 78 02 44 32 04 4C

This packet forces the AD2/DIO2 (pin 18) output to a logic 0, so no current will flow from the XBee module pin: Both sides of the LED see only a ground connection.

You also could use the following commands to turn off the LED.

D2 03, which changes the AD2/DIO2 pin to become a digital input

D2 02, which changes the AD2/DIO2 pin to connect as an input to the ADC

D2 00, which disables the AD2/DIO2 pin

Step 9. Create a packet that will turn on LED2 at the AD3-DIO3 pin, and another packet to turn it off:

LED On:

7E 00 10 17 52 00 00 00 00 00 00 00 00 56 78 02 ___ ___ ___ ___

LED Off:

7E 00 10 17 52 00 00 00 00 00 00 00 00 56 78 02 ___ ___ ___ ___

See the answers at the end of this experiment.

Unfortunately, the XBee modules do not have a command that lets you change the logic condition at several bits simultaneously. You must do so one at a time. Keep in mind, though, the API packets and AT commands used in this experiment all started with the same information:

7E 00 10 17 52 00 00 00 00 00 00 00 00 56 78 02 ___ ___ ___ ___

followed by a two-byte command, a parameter, and a checksum. You'll find similar patterns when you use API packets to transmit other AT commands. When a microcontroller sends an API packet, it simply sends one byte after another with no ASCII space values between them and usually no delay between them.

EXPERIMENT ANSWERS

Important: In this experiment you did not configure the two ADx-DIOx pins on the XMTR module as digital outputs. The API command performed this operation by causing the XMTR module to put these pins in a logic-1 or a logic-0 state. You could have used a command to cause a remote XBee module use an I/O pin as an analog input. The AT commands within an API packet give you many control options. You also could control PWM outputs at a remote XBee module, for example.

Step 6. Can you create a packet with a command to turn off LED1 at the AD2/DIO2 pin?

7E 00 10 17 52 00 00 00 00 00 00 00 00 56 78 02 44 32 04 4C

Step 9. Create a packet that will turn on LED2 and another packet to turn it off:

7E 00 10 17 52 00 00 00 00 00 00 00 00 56 78 02 44 33 05 4A

7E 00 10 17 52 00 00 00 00 00 00 00 00 56 78 02 44 33 04 4B

Remote Control of PWM Outputs

REQUIREMENTS

2 XBee modules
1 XBee adapter
3 LEDs
3 220-ohm resistors, 1/4 watt, 10% (red-red-brown)
1 Solderless breadboard
1 USB-to-XBee adapter
1 USB cable—type-A to mini-B
Digi X-CTU software running on a Windows PC, with an open USB port

INTRODUCTION

In this experiment you will learn how to use an application programming interface (API) command to control a PWM output on a remote XBee module without applying a voltage to the AD0-DIO0 or AD1-DIO1 input at a transmitting module. The PWM outputs have some particular traits you must pay careful attention to if you need to use PWM signals in a design. You will learn about those traits in this experiment.

Step 1. In this step you will configure the module labeled XMTR to operate with AT commands via the API. Ensure you have turned off power to the breadboard. Then place the XMTR module in the USB-to-XBee adapter and reconnect this adapter to the USB cable and your PC. You can find the configuration information that follows in the file EX13_XMTR.pro.

Within the X-CTU window, click on Modem Configuration and click on Restore. After a few seconds you should see the message "Restore Defaults.. complete" appear at the bottom of the X-CTU window. Click Read to obtain the default values from the XMTR module. Ensure the DL – Destination Address Low and MY – 16-Bit Source Address each have a value of 0. If they do not, perform Step 1 again.

The Hands-on XBee Lab Manual.

In the Modem Configuration window, set the following hex values for the XMTR module:

DL – Destination Address Low   = 1234
MY – 16-Bit Source Address   = 5678
AP – API Enable   = 1-API ENABLED
IA – I/O Input Address   = FFFF
P0 – PWM0 Configuration   = 2-PWM OUTPUT
PT – PWM Output Timeout   = FF

Save these configuration settings in the XMTR module and move it to its XBee adapter socket on the breadboard. Do not turn on power to the breadboard.

Step 2. If you just completed Experiment 12 and have not changed any of the Modem Configuration settings in the module labeled RCVR, please go to Step 3. Otherwise, place the RCVR module in the USB-to-XBee adapter and reconnect this adapter to the USB cable. You can find the configuration information that follows in file EX13_RCVR.pro.

Within the X-CTU window, click on Modem Configuration and click on Restore. After the message "Restore Defaults..complete" appears, click Read. If the DL – Destination Address Low and MY – 16-Bit Source Address are not equal to 0, perform Step 2 again.

In the Modem Configuration window, set the following values for the RCVR module:

DL – Destination Address Low   = 5678
MY – 16-Bit Source Address   = 1234
AP – API Enable   = 1-API ENABLED
IA – I/O Input Address   = FFFF

Save these configuration settings in the RCVR module and leave the RCVR module in the USB-to-XBee adapter socket connected to the PC.

Step 3. The circuit diagram in Figure 13.1 shows connections needed for the XMTR module in your breadboard. If you have connections other than +3.3-volts power (pin 1) and ground (pin 10) at the XMTR module, disconnect them now. One LED circuit connects to pin 6, the PWM0-RSSI output, and will indicate activity at this pin. The other two LEDs connect to the AD0-DIO0 and AD3-DIO3 pins to indicate their logic state. A logic 0 at a pin turns on the corresponding LED. I recommend you keep the PWM0 LED separate from the other two because you will monitor its operation most often.

Step 4. An XBee module has two PWM outputs, PWM0 and PWM1, with corresponding output-control commands M0 and M1. You also can set a PWM-output *configuration* via a P0 or P1 API command, but in this experiment you set the P0 configuration via the X-CTU Modem Configuration window in Step 1. A P0 or P1 command sent in an API packet would require the same settings available in the X-CTU Modem Configuration window: 0 – DISABLE, 1 – RSSI, or 2 – PWM OUTPUT.

Because you already set the PWM0 pin on the XMTR module to act as an output, you only need to send a PWM *value*, between 0X000 and 0x3FF, to control the pulse width. The output ranges from 0 to 100 percent, and it

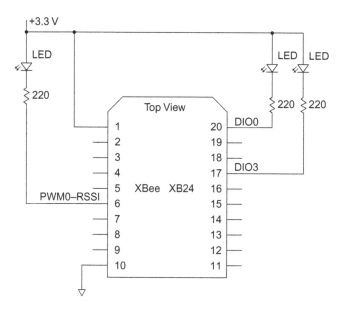

FIGURE 13.1 An LED connected to the PWM0-RSSI output lets you monitor activity of the PWM portion of an XBee module. Two other LEDs let you observe changes at the DIO0 and DIO3 pins.

provides 1024 values, from 0 through 1023_{10}. In the M0 API packet, the value 0x0380 will set the PWM0 output to about a 1-to-8 ratio of logic 0 to logic 1 periods, which will make the LED turn on about 12 percent of the time and appear dim. The higher the PWM value, the longer each pulse provides a logic-1 signal.

Now you will prepare to send an AT command to the remote XMTR module to control the PWM output. This experiment continues to use the 16-bit Source Address for the XMTR module instead of its 64-bit serial number. Thus your and my XMTR module have the same Source Address, 0x5678, which we set in the Modem Configuration window, so we can use the same API packet to control the PWM-output LED at the XMTR module. In the partial packet below, the 2-byte value that controls the PWM0 output follows the M0 command.

...

4D = hex code for letter M
30 = hex code for numeral 0
03 = most-significant byte of PWM value
80 = least-significant byte of PWM value

...

Important: When you transmit an M0 or an M1 command use the hex code for the letter M (0x4D) and the hex code for *numeral* 0 (0x30) or *numeral* 1 (0x31). Do not enter a 0 or 1. (The same requirements hold true for a P0 or

P1 command.) The M0 and M1 commands sent *without* the two data bytes let you read the value currently in use.

Step 5. If your breadboard is powered, remove power, wait about five seconds, and turn power on. If power is off to start, turn on power to your XMTR module now. The LED connected to the PWM0 output should turn on to full brightness because when reset a PWM output provides a logic-0 signal. Type the following packet into the Send Packet window but do not transmit it to the powered XMTR module.

```
7E 00 11 17 52 00 00 00 00 00 00 00 00 56 78 02 4D 30 03 80 C6
```

This experiment involves sending many command packets to the XMTR module, but most of the hex values remain the same. You need to change only the byte count value and the last few bytes in the command packets that follow. In the previous packet I underlined the bytes that change from packet to packet.

Please read this paragraph before you perform the steps it describes. Watch the LED and send the packet. The PWM0 LED should dim when the XMTR receives the M0 command packet. Continue to watch the LED for about 30 seconds. What did you observe? After the 30-second period, turn off power to the XMTR module and go back to the start of this paragraph, and time the period between the dimming of the LED and the next change in brightness.

The LED at my XMTR module dimmed and then about 22 seconds later it turned on to full brightness. Perhaps you thought the LED would remain dim due to a continuing PWM output of pulses.

Step 6. The Digi International manual for the XBee/XBee-PRO modules provides the following cryptic description of the PT – PWM Output Timeout, which you configured with the value 0xFF in Step 1:

PWM Output Timeout. Set/Read output timeout value for both PWM outputs. When PWM is set to a non-zero value: Due to I/O line passing, a time is started which when expired will set the PWM output to zero. The timer is reset when a valid I/O packet is received. Parameter Range: 0 – 0xFF [×100 msec]. (Ref. 1)

I assumed this description meant the PWM output would turn on with the pulse width set with the 0x380 value. Then, after the programmed period, which for 0xFF equals 25.5 seconds, the PWM output would become a logic 0. That matched the observation in Step 5, although I measured 22 seconds with a stopwatch.

The Digi description seems to imply the next I/O command would reset the timer and thus turn on the PWM output again. So I transmitted a command packet to change the DIO3 pin to a logic 0:

44 = hex code for letter D
33 = hex code for numeral 3
04 = 04 (DO-LOW)

Here's the complete packet for the AD3-DIO3 DO-LOW command:

```
7E 00 10 17 52 00 00 00 00 00 00 00 00 56 78 02 44 33 04 4B
```

Although this command turned on the LED connected to the DIO3 pin, it did not reactivate the PWM output. I also tried the command packet to set the AD0-DIO0 pin to a logic 0:

```
7E 00 10 17 52 00 00 00 00 00 00 00 00 56 78 02 44 30 04 4E
```

Likewise, this command failed to reactivate the PWM output, although it turned on the DIO0 LED. Try these command packets with your remote XBee module to confirm this behavior. Only the underlined values change.

Step 7. So how can you obtain a continuous output at the PWM0 pin? You might wonder if a *second* transmission of the M0 command packet would restart the PWM output. Turn off power to your XMTR module, wait a few seconds and turn power on again. This action resets the XMTR module. Then send the command packet for M0 with a value 0x0380:

```
7E 00 11 17 52 00 00 00 00 00 00 00 00 56 78 02 4D 30 03 80 C6
```

The PWM0 LED should dim and become brighter after 22 seconds when the PWM0 output changes to a logic 0. Resend the M0 command packet and observe the LED again. Does it brighten again after 22 seconds? In my lab, the second transmission of the packet caused the LED to dim and it remained that way. An oscilloscope confirmed the continuing presence of the PWM0 pulses. As long as the XMTR module remained powered, it produced the proper PWM signal. So it seems the first M0 command packet started the timer, but subsequent M0 command packets produce a continuing PWM output without the timed period. Do not turn off power to your breadboard or to the XMTR module.

Step 8. The Digi manual referred to earlier noted, "The timer is reset when a valid I/O packet is received." That statement seems to imply an I/O command will reset the timer and thus the PWM output. The next steps determine if this timer reset will occur and how I/O-port commands affect PWM operation.

Step 9. At this point, the LED on the PWM0 pin should remain dim because the PWM output continues to produce short logic-0 pulses. If the LED connected to the PWM0 pin on your XMTR module remains dim, skip ahead to Step 10. If the LED appears bright, or if in doubt, continue with this step.

Remove power from the XMTR module and breadboard, wait five-or-so seconds, and turn on power. Then send the packet that follows:

```
7E 00 11 17 52 00 00 00 00 00 00 00 00 56 78 02 4D 30 03 80 C6
```

Wait until the LED becomes bright again and send the packet a second time. The LED now should remain dim beyond the 22-second timeout period. That condition indicates the PWM0 output continues to produce short logic-0 pulses.

Step 10. Type the following command to set the DIO0 pin to a logic 0 and observe what—if anything—happens to the PWM LED brightness:

```
7E 00 10 17 52 00 00 00 00 00 00 00 00 56 78 02 44 30 04 4E
```

The LED gets brighter as soon as you send the command. Thus, the I/O command stopped the PWM output, which became a logic 0 again. Do not turn off power to your XMTR module. Resend the PWM M0 command packet you used earlier:

```
7E 00 11 17 52 00 00 00 00 00 00 00 00 56 78 02 4D 30 03 80 C6
```

Did the PWM output change the brightness of the LED? The LED should have dimmed, which means the PWM0 output again has a pulse output. The LED connected to the DIO0 pin should turn on. Now, resend the packet to again set the AD4-DIO4 pin to a logic 0:

```
7E 00 10 17 52 00 00 00 00 00 00 00 00 56 78 02 44 30 04 4E
```

Did the brightness of the PWM0 LED change? It should not change. Although you sent the AD0-DIO0 command twice, the first command changed the state of the pin from a logic 1 to a logic 0. The second command had no effect because the DIO0 output was already a logic 0. Thus, no *change* occurred at the DIO0 pin. You can conclude that only a change in the state of an I/O pin will stop the PWM outputs and force them to a logic-0 state.

When you consider this PWM-reset action and the reset caused by the PWM Output Timeout, maintaining a continuous PWM output becomes a chore. You can use a work-around, though, to get close to a continuous PWM output.

Step 11. Unfortunately, neither the Digi International Web site nor other sites provide examples of how to properly use the PWM Output Timeout or how to maintain a continuous PWM output, so I ran a few more experiments that you will duplicate in the following steps.

Turn off power to your breadboard, remove the XMTR module from its adapter socket, and place it in the USB-to-XBee adapter. Read its Modem Configuration information and find the PT – PWM Output Timeout setting under the I/O Line Passing heading. Set the PT value to zero: PT – PWM Output Timeout = 0.

Save this new configuration in the XMTR module and return it to the adapter in the breadboard, but do not turn on power. With a PWM timer value of 0, what might you expect?

In the X-CTU Send Packet window, re-enter the M0 PWM command packet:

```
7E 00 11 17 52 00 00 00 00 00 00 00 00 56 78 02 4D 30 03 80 C6
```

Turn on power to your breadboard and send the packet. The LED connected to the PWM0 output should dim and remain dim. No timeout occurs and the PWM0 output continues to produce the expected signal. After an hour or so, my XMTR module continued to produce the same PWM signal, as seen with an oscilloscope. During that time I had not sent the XMTR module any other commands. It seems as though a PT – PWM Output Timeout setting of 0 disables the timeout period for the PWM output, which now continues to produce pulses for as long at the XMTR module remains powered.

Step 12. Send the XMTR module the command packet used in Step 6 to cause the XMTR module to place a logic-0 signal on the DIO3 pin:

```
7E 00 10 17 52 00 00 00 00 00 00 00 00 56 78 02 44 33 04 4B
```

What happens to the PWM0 LED when you send this packet? That LED brightened, signaling the PWM0 output had reset to a logic 0. The LED connected to the DIO3 pin also turned on. Although setting the PT – PWM Output Timeout value to 0 eliminated the timeout period for the PMW0 signal, changing an I/O pin still stops the PWM signal and forces the PWM0 output to logic 0.

I found an imperfect solution to the problem, and, absent additional information from Digi or from someone with additional XBee-module PWM experience, it must suffice for a stand-alone XBee module. Because an I/O command might disable a PWM output I recommend you set PWM Output Timeout to 0 and follow any I/O command with a PWM M0 or M1 command to "refresh" the corresponding PWM value. Given the other useful operations of the XBee modules, Digi's approach to control of PWM outputs seems odd.

Step 13. Turn off power to your XMTR module and then turn on power. Send the XMTR module the command that follows to start the pulsed output on the PWM0 pin. You should see the PWM0 LED go dim.

```
7E 00 11 17 52 00 00 00 00 00 00 00 00 56 78 02 4D 30 03 80 C6
```

Then in the Send Packet window, type the following packet that includes two commands, placed in sequence:

```
7E 00 10 17 52 00 00 00 00 00 00 00 00 56 78 02 44 33 04 4B 7E
00 11 17 52 00 00 00 00 00 00 00 00 56 78 02 4D 30 03 80 C6
```

Do not press the Enter key, just keep typing so the Send Packet window has a continuous series of hex values. The first packet (underlined) forces the DIO3 pin to a logic 0 and the packet that immediately follows sends the same PWM command packet that created the PWM pulses at the start of this step. Watch the PWM0 LED and click Send Data. What did you observe on the LED?

The PWM0 LED produced a brief flash and then went back to its dim mode. To start, the PWM output produced the expected pulses that dimmed the LED. The command that turned on the DIO3 LED turned off the PWM pulses so the PWM0 pin dropped briefly to logic 0, which causes the LED to become bright. The second command in the packet turned the PWM back on with the original value, 0x380, so the LED went back to its dim condition. The bright flash appeared for about 17 msec. Figure 13.2 provides a timing diagram that shows the relationship of the PWM0 output to the DIO3 output change.

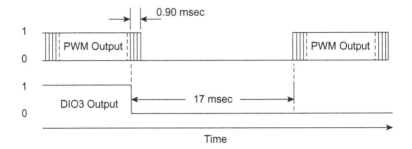

FIGURE 13.2 This timing diagram shows the relationship between the change at an output pin caused by a command packet and the end of a series of PWM pulses. A second command packet reloads the PWMO value and the PWM output restarts.

Step 14. Resend the long packet that contains the DIO3 and the PWM command to the XMTR module, what do you think happened? Nothing happens, because the DIO3 command does not *change* any of the I/O pin settings. Only a change turns off the PWM output.

A BETTER APPROACH

Given the difficulty sorting out the use of the PWM capabilities in an XBee module, if you need a continuous PWM output without the glitches introduced by I/O changes I recommend using a small microcontroller. Many MCUs provide a PWM output that operates without interruption. An XBee module could send PWM values to an attached MCU via its serial port. Likewise, the MCU could provide many types of I/O ports not available on an XBee module. In this situation, let the XBee module provide wireless communications and have the MCU control analog, digital, and PWM signals.

REFERENCE

"XBee/XBee-PRO RF Modules," Product Manual V1.xEx – 802.15.4 Protocol, Digi International. 2009.

How to Parse Data from Digital and Analog Pins and Control Individual Digital Outputs

REQUIREMENTS

2 XBee modules
1 XBee adapter board
1 220-ohm resistor, 1/4 watt, 10% resistor (red-red-brown)
1 4700-kohm, 1/4W, 10% resistor (yellow-violet-red)
1 10-kohm, 1/4W, 10% resistor (brown-black-orange)
1 10-kohm potentiometer
1 LED
1 Solderless breadboard
1 USB-to-XBee adapter
1 USB cable—type-A to mini-B
Digi X-CTU software running on a Windows PC, with an open USB port

INTRODUCTION

In this experiment you will learn how to use an application programming interface (API) command to ask a remote XBee module to reply with information about the state of its digital I/O pins *and* the values from all active analog-to-digital converter (ADC) inputs.

Step 1. In this step you will configure the remote XMTR module to operate with the AT commands via API packets. Ensure you have turned off power to the breadboard. Then place the XMTR module in the USB-to-XBee adapter and reconnect this adapter to the USB cable and your PC.

Within the X-CTU window, click on Modem Configuration and then click on Restore. After a few seconds you should see the message "Restore

The Hands-on XBee Lab Manual.

Defaults..complete" appear. Click on Read to obtain the restored default values from the XMTR module.

Check to ensure the DL – Destination Address Low and MY – 16-Bit Source Address each have a value of 0. This condition indicates the X-CTU software has reset all the XMTR module settings to their default state. If DL and MY are not equal to 0, please retry the procedure in this step.

In the Modem Configuration window, set the following hex values for the XMTR module. You can find this configuration information in the file: EX14_XMTR.pro.

DL – Destination Address Low	= 1234
MY – 16-Bit Source Address	= 5678
AP – API Enable	= 1-API ENABLED
IA – I/O Input Address	= FFFF
D7 – DIO7 Configuration	= 3-DI
D6 – DIO6 Configuration	= 3-DI
D5 – DIO5 Configuration	= 2-ADC
D4 – DIO4 Configuration	= 2-ADC
D3 – DIO3 Configuration	= DO LOW
D2 – DIO2 Configuration	= DO LOW
D1 – DIO1 Configuration	= DO HIGH
D0 – DIO0 Configuration	= DO HIGH

These settings establish two digital-input pins (D7 and D6), two ADC-input pins (D5 and D4), two digital outputs set at logic 0 (D3 and D2), and two digital outputs set at logic 1 (D1 and D0). Double check the configurations above and save them in the XMTR module.

Remove the USB-to-XBee adapter from the USB cable and remove the XMTR module from the adapter socket. Place the XMTR module in its XBee adapter socket on the breadboard.

Step 2. Place the RCVR module in the USB-to-XBee adapter and reconnect this adapter to the USB cable.

Within the X-CTU window, click on Modem Configuration and then click Restore. After a few seconds you should see the message "Restore Defaults.. complete" appear. Click on Read to obtain the restored default values from the RCVR module.

Check to ensure the DL – Destination Address Low and MY – 16-Bit Source Address each have a value of 0. This condition indicates the X-CTU software has reset all the RCVR module settings to their default state. If DL and MY are not equal to 0, please perform this step again.

In the Modem Configuration window, set the values shown next for the RCVR module. You can find this configuration information in the file: EX14_RCVR.pro.

DL – Destination Address Low	= 5678
MY – 16-Bit Source Address	= 1234
AP – API Enable	= 1-API ENABLED
IA – I/O Input Address	= FFFF

Save these configuration settings in the RCVR module. Leave the RCVR module in the USB-to-XBee adapter socket connected to the PC.

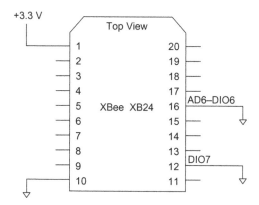

FIGURE 14.1 Connections needed to create a logic-0 signal for two XBee inputs.

Step 3. The schematic diagram in Figure 14.1 shows the connections needed for the XMTR module in your breadboard. If you have other connections at the XMTR module, please disconnect them now and make only the connections shown in Figure 14.1.

Step 4. This experiment continues to use a remote module's 16-bit address, so your and my XMTR module have a Source Address of 0x5678, which we set in the Modem Configuration window and saved in our XMTR modules.

The 2-character AT command IS – Force Sample causes the addressed module to reply with information about all its I/O ports and active ADC inputs. This command has no parameters associated with it. The characters I and S have equivalent ASCII hex values: I = 0x49 and S = 0x53.

The partial packet below includes the IS command:

```
...
5678    = 16-bit source address (network address)
02      = Value that causes immediate actions
49      = Hex code for letter I
53      = Hex code for letter S
??      = Checksum
```

Step 5. Turn on power to your XMTR module in the breadboard and switch to the X-CTU Terminal window. Clear the Terminal window and ensure you have it set for Show Hex. Click on Assemble Packet and in the Send Packet window, type the following hex values:

7E 00 0F 17 52 00 00 00 00 00 00 00 00 56 78 02 49 53 2A

Click Send Data. You should see hex values as shown in Figure 14.2, but not all these values will match yours. The information that follows shows the portion of the reply that follows the IS command and the status byte:

```
...
01      = one sample of I/O lines and ADC inputs
60      = first active-signal byte
CF      = second active-signal byte
00      = first digital-data byte
```

03   = second digital-data byte
03FF   = analog sample from AD4-DIO4
03FF   = analog sample from AD5-DIO5

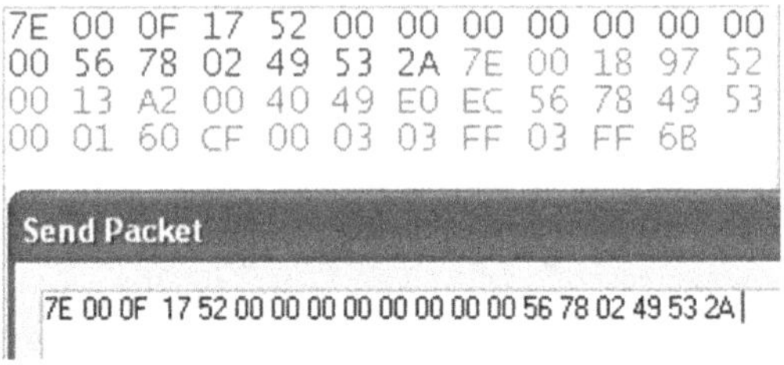

```
7E 00 0F 17 52 00 00 00 00 00 00 00
00 56 78 02 49 53 2A 7E 00 18 97 52
00 13 A2 00 40 49 E0 EC 56 78 49 53
00 01 60 CF 00 03 03 FF 03 FF 6B
```

Send Packet

```
7E 00 0F  17 52 00 00 00 00 00 00 00 00 56 78 02 49 53 2A|
```

FIGURE 14.2 The X-CTU terminal shows the results from sending the IS command.

You set the XMTR module for D5 and D4 as ADC inputs, so the 0x60 byte identifies these active inputs as shown in Table 14.1. You did not configure the D8 pin as an input, so it remains 0; inactive. The second active-signal byte, 0xCF, also shown in Table 14.1, indicates all pins set for digital inputs *except* for the AD5-DIO5 and AD4-DIO4, which now serve as analog inputs.

Table 14.1 The Active-Signal Bytes in the Reply to an IS Command

	First Active-Signal Byte							
	First Hex Character				Second Hex Character			
Bit Position	B7	B6	B5	B4	B3	B2	B1	B0
Bit Function	X	A5	A4	A3	A2	A1	A0	D8
Binary Data	0	1	1	0	0	0	0	0
Hex Data		6				0		
	Second Active-Signal Byte							
	First Hex Character				Second Hex Character			
Bit Position	B7	B6	B5	B4	B3	B2	B1	B0
Bit Function	D7	D6	D5	D4	D3	D2	D1	D0
Binary Data	1	1	0	0	1	1	1	1
Hex Data		C				F		

The next two bytes, 0x00 and 0x03, shown in Table 14.2, give you the state of the digital pins. You can ignore the first digital-data byte because the D8 input was not set for any digital input.

For the second digital-data byte 0x03, also shown in Table 14.2, you see a logic 1 for D1 and D0 because you configured them as digital-output-high (DO-HIGH) pins. Pins D3 and D2 show logic 0 because you configured them as digital-output-low (DO-LOW) pins. You grounded input pins D7 and

D6 on the breadboard for the XMTR module, so they appear as logic 0. You can ignore the D5 and D4 bits (grey background) because they correspond to analog inputs.

Table 14.2 The Digital-Data Bytes in the Reply to an IS Command

	First Digital-Data Byte							
	First Hex Character				Second Hex Character			
Bit Position	B7	B6	B5	B4	B3	B2	B1	B0
Bit Function	X	X	X	X	X	X	X	D8
Binary Data	0	0	0	0	0	0	0	0
Hex Data		0				0		

	Second Digital-Data Byte							
	First Hex Character				Second Hex Character			
Bit Position	B7	B6	B5	B4	B3	B2	B1	B0
Bit Function	D7	D6	D5	D4	D3	D2	D1	D0
Data	0	0	0	0	0	0	1	1
Hex Data		0				3		

The next two hex values, 0x03FF and 0x03FF, represent the voltages present at the ADC inputs you selected, D4 and D5. Because you did not connect these inputs to an external signal, they "float" to a 3.3-volt level and give a full-scale 10-bit value of 0x3FF.

Important: The IS - Force Sample command returns information not only for active inputs, but also for the state of any active *outputs*. This command gives you a way to quickly determine what data—input or output—exists on the digital I/O pins and it forces an immediate analog-to-digital conversion of the voltages present on any active ADC inputs. Later when you work with an XBee module and an MCU you will learn how to use the IS command in software to poll all active XBee modules to gather information from them. The IS command can play an important part in the repertoire of commands used in a network of modules that collect information from remote devices.

As you can guess, creating an API command packet for each module can become tedious and could lead to errors in inserting proper addresses. Thankfully, MCU software can handle these tasks. Creating packets by hand only makes sense for testing and experiments such as those in this book.

Step 6. In this step you will change the inputs at four pins and add an LED as an output indicator, as described next and shown in Figure 14.3.

• Turn off power to your XMTR module
• At DIO7 (pin 12) and DIO6 (pin16) change the logic-0 (ground) inputs to logic-1 (+3.3 volts) inputs.

- Connect the a 10-kohm potentiometer between +3.3 volts and ground and connect the potentiometer's center contact to the AD5-DIO5 pin (pin 15). Adjust the potentiometer to about halfway between its end points.
- Connect a 10-kohm fixed resistor (brown-black-orange) and a 4.7-kohm resistor (yellow-violet-red) as shown in Figure 14.3 and connect the junction of the resistors to the AD4-DIO4 pin (pin 11).
- Connect the XBee module ADC reference line, VREF (pin 14) to +3.3 volts.
- Connect the LED to the AD2-DIO2 pin as shown.

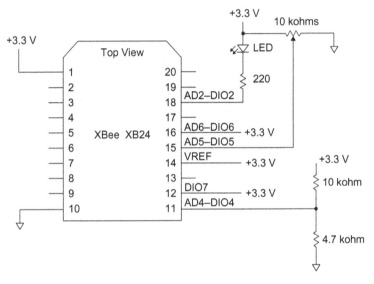

FIGURE 14.3 Connections for analog and digital inputs at the remote XMTR module to provide voltages to the ADC inputs and logic levels for digital inputs.

Step 7. Turn on power to the breadboard and XMTR module and the LED should turn on because the AD2-DIO2 pin at the XMTR module still has the DO LOW setting established in Step 1. Thus current flows through the LED and into the XBee module.

Open the X-CTU Terminal window and clear it. In the Send Packet window you should still see the packet entered in Step 5. If not, go to Step 5 and re-enter the packet. Click on Send Data. What do you see in the Terminal window? I saw the information shown in Figure 14.4. Our XMTR-module serial-number bytes will not match.

Refer to Tables 14.1 and 14.2 or use blank tables in Appendix I to help you interpret the I/O-pin data obtained from your XMTR module. The data I received follows and your data should look similar.

01	= 1 sample of I/O lines and ADC inputs
60	= D5 and D4 set as analog inputs
CF	= D7, D6, and D3--D0 set as digital pins
00	= First digital-data byte (D8)

C3 = Second digital data byte, D7, D6, D1 D0 = logic 1 and D2 and
    D3 = logic 0

02B8 = 10-bit analog value from AD4 (will vary from my data)

0211 = 10-bit analog value from AD5 (will vary from my data)

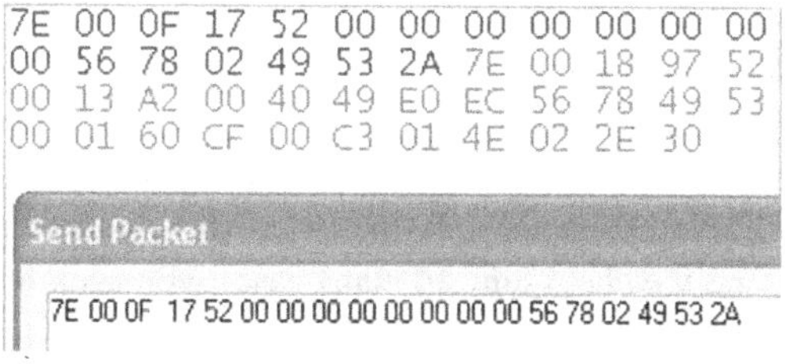

```
7E 00 0F 17 52 00 00 00 00 00 00 00
00 56 78 02 49 53 2A 7E 00 18 97 52
00 13 A2 00 40 49 E0 EC 56 78 49 53
00 01 60 CF 00 C3 01 4E 02 2E 30
```

Send Packet

7E 00 0F 17 52 00 00 00 00 00 00 00 00 56 78 02 49 53 2A

FIGURE 14.4 This information shows the effect of adding digital and analog signals at the XMTR module and sending the IS command.

Step 8. Change the potentiometer setting, resend the IS-command packet, and confirm the analog value from input-pin DIO5 changes. You can change the logic level at the digital inputs DIO7 (pin 12) and DIO6 (pin 16) and confirm they change, too.

Now you know how you can force a module to transmit the state of its digital inputs and outputs as well as perform an analog-to-digital conversion at active ADC inputs, on command. Note, though, you cannot force this pin-sampling action when you have a module in either the DOZE or HIBERNATE sleep mode.

Step 9. Could you create a command packet to turn the LED off? What state do you need at the AD2-DIO2 pin to turn off the LED?

Use the following "framework" to create a command packet that turns the LED off:

7E 00 __ 17 52 00 00 00 00 00 00 00 00 56 78 02 __ __ __ __ __ __ __ __

The underlined spaces leave room for the command, parameter, byte-count, and checksum values you think appropriate. Hint: Look in the X-CTU Modem Configuration window for the setting that configures the D2 pin. Find an answer at the end of this experiment.

After you turn the LED off, re-enter the command packet used earlier to force a sample of the I/O pins at the XMTR module:

7E 00 11 17 52 00 00 00 00 00 00 00 00 56 78 02 49 53 2A

Can you determine from the reply to this command that the AD2-DIO2 pin has a logic-1 (5 – DO HIGH) setting?

AT commands can manipulate individual bits at a remote XBee module, an important capability when you need to control individual devices without disturbing others. You could control sprinkler valves, door locks, automobile ignition, appliances, motors, and so on. The reply to an IS command lets you confirm settings so you can determine the state of inputs and outputs at a remote module as well as sample analog signals.

EXPERIMENT ANSWERS

Step 8. You can use the AT command D2 to control the AD2-DIO2 pin. To force that pin to a logic-1 state, use the command D2 05, which configures the pin for a data-output-high (DO – HIGH) state. The packet shown next will turn the LED off:

7E 00 <u>10</u> 17 52 00 00 00 00 00 00 00 00 56 78 02 <u>44</u> <u>32</u> <u>05</u> <u>4B</u>

To turn the LED back on, use D2 04 to configure the pin for a data-output-low (DO – LOW) state.

7E 00 <u>10</u> 17 52 00 00 00 00 00 00 00 00 56 78 02 <u>44</u> <u>32</u> <u>04</u> <u>4C</u>

In each packet, you changed only the byte that set a configuration for the selected I/O pin and the checksum. In many cases, you need not create a completely new packet when you work with I/O pins.

How to Control Several XBee Modules with Broadcast Commands

REQUIREMENTS

3 XBee modules
2 XBee adapter boards
4 LEDs
4 220-ohm, 1/4 watt, 10% resistors (red-red-brown)
1 Solderless breadboard
1 USB-to-XBee adapter
1 USB cable—type-A to mini-B
Digi X-CTU software running on a Windows PC, with an open USB port

INTRODUCTION

In this experiment you will learn how to communicate with more than one remote XBee module via wireless communications. You will use the application programming interface (API) to send AT commands to two remote modules to turn LEDs on or off. In previous experiments you used two modules, one marked XMTR and one marked RCVR. This experiment adds a third XBee module, which I labeled END. If you have only two XBee modules, you can skip this experiment, but I recommend you read at least through it.

You also will learn how to use a broadcast command that affects all remote modules and how to change XBee parameters semi-permanently so they remain set when power ceases to an XBee module.

Step 1. In this step you configure the XMTR module to operate with the AT commands via the API. Ensure you have turned off power to the breadboard. Then place the XMTR module in the USB-to-XBee adapter and reconnect this adapter to the USB cable and your PC.

Within the X-CTU window, click on Modem Configuration and then click Restore. After a few seconds you should see the message "Restore Defaults.. complete" appear. Click on Read to obtain the restored default values from the XMTR module.

Check to ensure the DL – Destination Address Low and MY – 16-Bit Source Address each have a value of 0. This condition indicates the X-CTU software has reset all the XMTR module settings to their default state. If DL and MY are not equal to 0, retry the procedure in this step.

In the Modem Configuration window, set the following hex values for the XMTR module. Find this configuration information in the file: EX15_XMTR.pro.

DL – Destination Address Low = 1234
MY – 16-Bit Source Address = 5678
AP – API Enable = 1-API ENABLED
IA – I/O Input Address = FFFF
D3 – DIO3 Configuration = 4 – DO LOW
D1 – DIO1 Configuration = 5 – DO HIGH

These settings establish two digital-output pins on the XMTR module, one set to a logic 0 and the other set to a logic 1. Double check the configurations above and save them in the XMTR module.

Remove the USB-to-XBee adapter from the USB cable and remove the XMTR module from the adapter socket. Insert the XMTR module in its XBee adapter socket on the breadboard.

Step 2. For a second XBee module you also will use as a remote device, follow the procedure in Step 1, but in the Modem Configuration window, set the following hex values for this module, which in my lab I labeled END. Find this configuration information in the file: EX15_END.pro.

DL – Destination Address Low = 0
MY – 16-Bit Source Address = 89FA
AP – API Enable = 1-API ENABLED
IA – I/O Input Address = FFFF
D3 – DIO3 Configuration = 4 – DO LOW
D1 – DIO1 Configuration = 5 – DO HIGH

These settings establish two digital-output pins on the END module, one set to a logic 0 and the other set to a logic 1. Note: The END module has a 16-bit Source Address of 89FA. Double check the configurations above and save them in the END module.

Remove the USB-to-XBee adapter from the USB cable and remove the END module from the adapter socket. Insert a second XBee adapter in your solderless breadboard about 3 inches (7.5 cm) from the adapter used for the XMTR module. Insert the END module into this second adapter. You should now have two XBee modules, XMTR and END, in XBee adapters on your breadboard.

Step 3. Place the RCVR module in the USB-to-XBee adapter and reconnect this adapter to the USB cable.

Within the X-CTU window, click on Modem Configuration and then click Restore. After a few seconds you should see the message "Restore

Defaults..complete" appear. Click on Read to obtain the restored default values from the RCVR module.

Check to ensure the DL – Destination Address Low and MY – 16-Bit Source Address each have a value of 0. This condition indicates the X-CTU software has reset all the RCVR module settings to their default state. If DL and MY are not equal to 0, perform Step 3 again.

In the Modem Configuration window, set the following values for the RCVR module. Find this configuration information in the file: EX15_RCVR.pro.

DL – Destination Address Low   = 5678
MY – 16-Bit Source Address   = 1234
AP – API Enable   = 1-API ENABLED
IA – I/O Input Address   = FFFF

Check your configuration settings. Save these configuration settings in the RCVR module. Leave the RCVR module in the USB-to-XBee adapter socket connected to the PC.

Step 4. The schematic diagram in Figure 15.1 shows the connections needed for both the XMTR and the END module in your breadboard. If you have other wires or components attached to the XMTR or the END module, please disconnect them now. Then add the components shown in Figure 15.1.

The photo in Figure 15.2 shows the arrangement of the XBee modules and LEDs on my breadboard. I suggest you use a similar arrangement—or otherwise label the LEDs—so you know which one corresponds to each output on a module.

Step 5. Turn on power to your breadboard and the two XBee modules, XMTR and END. LED1 connected to the AD1-DIO1 pin on each module should remain off and LED3 connected to the AD3-DIO3 pin should turn on. Do you know why? (See the Answers section at the end of this experiment.)

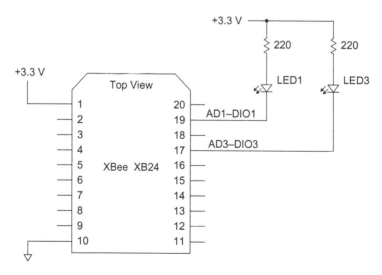

FIGURE 15.1 This schematic diagram applies to the XMTR and the END XBee module used in this experiment. (There is no LED2.)

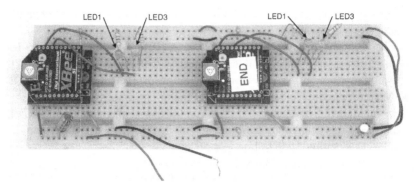

FIGURE 15.2 Keep the LEDs near their respective XBee module and make sure you know which LED corresponds to the AD1-DIO1 or AD3-DIO3 pin.

Step 6. Now you will write commands to change the state of the LEDs. The following API command packets assume 16-bit addressing; 5678 for the XMTR, and 89FA for the END modules. I have completed the first command and leave it to you to create the other three. You need change only the underlined bytes to address a different module and change the state of either the AD1-DIO1 or AD3-DIO3 pin. Remember, to turn an LED off, you need a logic 1 at the corresponding output. A logic 0 will turn on the associated LED. If you get stuck, see Answers at the end of this experiment.

- Turn off LED3 at XMTR module:
  7E 00 10 17 52 00 00 00 00 00 00 00 00 <u>56 78</u> 02 <u>44 33 05 4A</u>
- Turn on LED1 at XMTR module:
  7E 00 10 17 52 00 00 00 00 00 00 00 00 ___ ___ 02 ___ ___ ___ ___
- Turn off LED3 at END module:
  7E 00 10 17 52 00 00 00 00 00 00 00 00 ___ ___ 02 ___ ___ ___ ___
- Turn on LED1 at END module:
  7E 00 10 17 52 00 00 00 00 00 00 00 00 ___ ___ 02 ___ ___ ___ ___

Step 7. Note the state of the LEDs after you have sent the commands above and the LEDs have changed on/off conditions.

XMTR: LED1 _____ LED3 _____ END: LED1 _____ LED3 _____

Now turn off the power to the XMTR and END modules, wait a few seconds, and turn the power on again. What happens to the state of the LEDs? Do you know why? For an explanation, go to the Answers section at the end of this experiment.

Step 8. You can use the WR-Write command to change a remote module's configuration and make those changes remain constant even when you turn off power to a module. In this step you will use the WR command to change the power-up reset condition for the AD1-DIO1 configuration at the XMTR module.

If not already powered, turn on power to your breadboard. Each XBee module should have its LED3 lit. The LED1 LEDs should remain off. Use the command you created earlier to turn on LED1 at the XMTR module (0x5678).

If you do not have a command packet written down, see the answer for Step 6 at the end of this experiment.

Now both LED1 and LED3 at the XMTR module should be lit. To make this configuration remain as now set during power-up reset of the XMTR module, you must send a WR command. This command requires no parameters and it has one byte fewer than the command packets sent to remote modules to control LEDs:

7E 00 0F 17 52 00 00 00 00 00 00 00 00 56 78 02 57 52 1D

The packet above includes the byte 0x02 for immediate action, followed by the ASCII values for the letters W (0x57) and R (0x52). Go to the Send Packet window, type this packet, and send it. (Did you remember to change the third byte that indicates packet length?)

What did you observe? You should see only a reply in the X-CTU Terminal window with a status byte that indicates "OK." The WR-Write command did not affect the LEDs.

Turn off power to your breadboard, wait a few seconds, and reapply power. What do you observe now? The LED1 and LED3 at the XMTR module turn on because you have modified the I/O-pin configuration and saved it in the XMTR module.

Now you must change the configuration for the XMTR module to its original state for this experiment. You could send the module two packets to do this, but I recommend repeating Step 1, which uses the X-CTU program. When you do, first read the configuration from the module and look at the I/O Settings list. You should see both the D3-DIO3 and D1-DIO1 configurations set for 4 - DO-LOW. Now please proceed with Step 1 and then go on to Step 9.

Step 9. This experiment with three XBee modules created a peer-to-peer (equal-to-equal) network because all three XBee modules could communicate with each other. In this case, the RCVR module communicated with the XMTR and END modules, but the XMTR and END modules also could communicate with each other by using their respective 16-bit network addresses and setting their high and low addresses to zero in commands. This experimental setup, though, doesn't provide an easy way to initiate a transmission by the XMTR or END modules.

If you have only a few modules, this type of peer-to-peer network might work well, but if you have five or more transceivers, without careful planning, you could end up with communication chaos. Imagine an unruly meeting in which several groups have separate conversations going on. A later experiment will have you set up a network of end-device modules and one coordinator, which simplifies control.

Step 10. The XBee modules let you transmit a *broadcast* message or command that all modules within range receive simultaneously. The remote modules also acknowledge a broadcast message. So you can determine which modules received the broadcast and which did not. A broadcast message could help you synchronize all modules, send a timing signal to all modules, turn off all devices immediately, and so on.

The following information explains how to set up and send a broad-cast packet, which follows the same general format used with other packets. A complete broadcast packet that causes all modules to place their AD1-DIO1 pin in a logic-0 state looks like this:

7E	= start byte
0010	= number of bytes in transmission
17	= remote AT-command-request byte
52	= frame identifier value (all experiments use this value)
000000000000FFFF	= 64-bit *broadcast* identifier
FFFE	= value needed to force 64-bit addressing above
02	= value that causes immediate actions
44	= hex code for D
31	= hex code for numeral 1
04	= hex code for DO – LOW
20	= checksum

The 0xFFFE value used as the 16-bit network address directs modules to use the transmitted 64-bit address instead. For XBee modules, the address 0x000000000000FFFF indicates they all must respond to this message regard-less of their factory-programmed serial number. The broadcast message will turn on LED1 on the XMTR and END modules simultaneously.

Turn off power to the breadboard and XMTR and END modules. Turn power on again. You should see LED3 lit at each module. Type the command above into the Send Packet and click on Send Data. (Remember, this packet has a byte count of 0x10.) You should see LED1 turn on at each module.

Now complete the following packet to broadcast a command to the XMTR and END modules to turn off LED3:

7E 00 10 17 52 00 00 00 00 00 00 FF FF FF FE 02 ___ ___ ___ ___

Find a complete packet in the Answers section at the end of this experiment.

Step 11. Each module—END and XMTR—responded to the broadcast message with its own acknowledgement, as shown in the information that follows. (Your XBee-module serial number or numbers will not match mine, which I underlined.)

7E 00 0F 97 52 <u>00 13 A2 00 40 49 E1 E6</u> 89 FA 44 33 00 17
7E 00 0F 97 52 <u>00 13 A2 00 40 49 E0 EC</u> 56 78 44 33 00 C7

Both responses include standard information: each module's serial number, network address, the D3 command, a status byte, and a checksum. If you had additional modules, they also would send an acknowledgement message.

Although you can send a broadcast packet, use it with care. Unless you need simultaneous actions, you might get better control by sending a message to individual modules and then monitoring for an acknowledgement from each before transmitting to another module.

EXPERIMENT ANSWERS

Step 5. When you apply power to a module, it starts in the configuration pro-grammed with the X-CTU software. If you use a command to change the logic level at an output pin, for example, it remains "volatile." That means when you turn off power and re-power a module, it reverts to the configurations set with the X-CTU software. In this experiment you will learn how to save new configuration settings in remote modules.

Step 6.

- Turn off LED3 at XMTR module:

  7E 00 10 17 52 00 00 00 00 00 00 00 00 56 78 02 44 33 05 4A

  Use a D3 command for LED3 with the value 04 for a logic 1 at the AD3-DIO3 pin at the XMTR module (0x5678).

- Turn on LED1 at XMTR module:

  7E 00 10 17 52 00 00 00 00 00 00 00 00 56 78 02 44 31 04 4D

  Use a D1 command for LED1 with the value 05 for a logic 0 at the AD1-DIO1 pin at the XMTR module (0x5678).

- Turn off LED3 at END module:

  7E 00 10 17 52 00 00 00 00 00 00 00 00 89 FA 02 44 33 05 95

  Use a D3 command for LED3 with the value 04 for a logic 1 at the AD3-DIO3 pin at the END module (0x89FA).

- Turn on LED1 at END module:

  7E 00 10 17 52 00 00 00 00 00 00 00 00 89 FA 02 44 31 04 98

  Use a D1 command for LED1 with the value 05 for a logic 0 at the AD1-DIO1 pin at the END module (0x89FA).

Step 7. The LEDs at the XMTR and END modules return to the original conditions set with the X-CTU software. The LED connected to the AD3-DIO3 pin at each module turns on, and the other LED turns off. This situation reflects the configuration you set in each module in Steps 1 and 2. Even though you changed the state of the AD1-DIO1 and AD3-DIO3 pins at each module, those changes exist only as long as you maintain power to a module. When a module goes through a power-up reset, it reverts to the original settings, so you know any attached equipment starts in a known state.

Step 10. Now complete the following packet to broadcast a command to the XMTR and END modules to turn off LED3:

  7E 00 10 17 52 00 00 00 00 00 00 FF FF FF FE 02 44 33 05 1D

How to Communicate Between an MCU and an XBee Module

REQUIREMENTS

2 XBee modules
1 XBee adapter board
1 Solderless breadboard
1 USB-to-XBee adapter
1 USB cable—type-A to mini-B
1 Microcontroller with a serial port (Arduino Uno or ARM mbed)
1 5-V-to-3.3-V logic-conversion circuit or module (for Arduino Uno)
Terminal-emulation software such as HyperTerminal for Windows
Digi X-CTU software running on a Windows PC, with an open USB port

INTRODUCTION

In this experiment you will learn how to transfer information from a micro-controller to a PC by using one XBee module as a transmitter and another as a receiver. You can send as many bytes as you need, and they can include typed information, data from measurements, error and status messages, and so on. The type of information depends solely on your requirements. As such, the XBee transmitter and receiver modules acts as "transparent" devices as though they have a direct, wired connection. The wireless communications neither include nor require any commands for XBee modules.

This experiment uses your XMTR module with a microcontroller (MCU) to create information the XMTR will transmit to the RCVR module. The RCVR module connects to your PC, which will display results in a terminal-emulator window. The diagram in Figure 16.1 shows the equipment setup.

The Hands-on XBee Lab Manual.

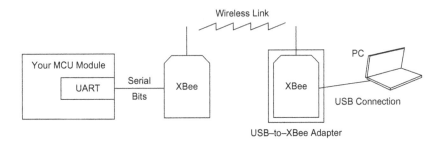

FIGURE 16.1 A block diagram of microcontroller-to-computer communications with XBee modules.

This experiment works directly with either an ARM mbed module or an Arduino Uno module, although I encourage experiments with other MCUs and boards on your own.

Several vendors offer these MCU modules, peripheral-device boards, and add-on devices at low cost and they provide easy ways to run experiments without getting deeply into hardware or software details. See Appendix D for information about where to obtain these products.

Step 1. In this step you will configure the XMTR module to operate with its default factory settings. Ensure you have turned off power to the breadboard. Then remove the XMTR module from the XBee socket in your breadboard and place it in the USB-to-XBee adapter and reconnect this adapter to the USB cable and your PC.

If not already running, start the X-CTU software. Within the X-CTU window, click on Modem Configuration and then Restore. After a few seconds you should see the message "Restore Defaults..complete" appear. Click on Read to obtain the restored default values from the XMTR module. Check to ensure the DL – Destination Address Low and MY – 16-Bit Source Address each have a value of 0. This condition indicates the X-CTU software has reset all XMTR module settings to their default state. If DL and MY are not equal to 0, retry the procedure in this step. After you confirm the default condition, place the XMTR module back in its adapter on the solderless breadboard.

Step 2. Place the RCVR module in the USB-to-XBee adapter and reconnect this adapter to the USB cable. Perform the configuration described in the second paragraph of Step 1. After you have restored the default settings for the RCVR XBee module, leave it in the USB-to-XBee adapter connected to your PC.

Step 3. The schematic diagram in Figure 16.2 shows the connections needed for the XMTR module in your breadboard. You need only power, ground, and a connection to your MCU. You may connect the wire to pin 3 on the XMTR module now, but do not connect it to your MCU. If you have other components or wires connected to the XMTR module, please disconnect them now. You also need a ground connection between your MCU and the XBee breadboard.

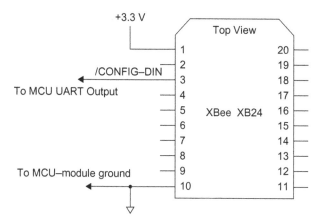

FIGURE 16.2 Connections at the XBee XMTR module for Experiment 16.

In the next steps, you will connect a serial-output pin on an MCU to the XMTR module. The Arduino Uno provides 5-volt logic signals, and the ARM mbed module provides 3.3-volt logic signals. The XBee modules work with 3.3-volt signals, so you cannot connect an XBee module directly to an Arduino Uno.

See Appendix A for instructions that describe how to build a simple, inexpensive logic-level converter, a schematic diagram and bill of materials. You need this circuit only if you plan to use a 5-volt MCU such as the Arduino Uno. The ARM mbed does not require logic-level conversions, nor do other 3.3-volt MCUs.

Important: This experiment assumes you already know how to program and use an Arduino Uno or ARM mbed MCU module, or that you're learning how to use one of them. The following steps include software examples for each MCU, but they do not provide step-by-step information about how to use the corresponding software tools that work with the ARM mbed or Arduino Uno modules. If you need tutorial information, visit the Arduino and mbed Web sites. Each module has its own group of supporters so you can find useful "how to" information, code, and examples on the Internet.

Experiments that use an MCU board will provide complete listings and code, so you will not need to write programs. Download all the code from: http://www.elsevierdirect.com/companion.jsp?ISBN=9780123914040.

USING A UART

Experiment 10 provided a short introduction to serial communications; the information that follows offers more details. If you have experience with serial devices, feel free to skip this section. Almost every MCU provides at least one Universal Asynchronous Transmitter/Receiver (UART) interface that will send and receive streams of logic bits via a separate transmit output and receive input. A UART takes a byte you supply, inserts a logic-0 start bit at the beginning of its "packet" and ends with one or two logic-1 stop bits, as shown in Figure 16.3. When you configure a UART via software, you specify its bit rate, the number of data bits in a transmission (almost always 8), the number of stop bits (usually only one), and parity (usually none). The parity bit can help detect errors in communications, but for the most part, we don't use it. Most MCUs offer standard data rates such as 1200, 2400, 9600, and 115,200 bits/ second.

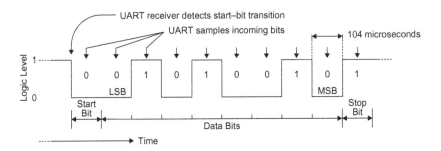

FIGURE 16.3 The stream of bits for a 9600 bits/sec. transmission of 0x4A.

The word "asynchronous" indicates UARTs in separate circuits do not share a synchronizing, or clocking, signal to time the data transfers. Instead, a UART receiver monitors its input and recognizes a logic-1-to-logic-0 transition as the beginning of a start bit (always logic 0). Then it acquires the following eight data bits and a parity bit (if any). The transmission ends when the UART receives a valid stop bit (always a logic 1). Bits always have equal duration and the UART samples them at the middle of their time slot as shown by the small arrows in Figure 16.3. Each bit in a 9600 bits/second transmission, for example, has a period of 104 microseconds. To operate properly, the receiving and transmitting UARTs must have the same data rate and the same configuration for the number of data bits, number of stop bits, and parity.

You might wonder why a UART doesn't see every logic-1-to-logic-0 transition as the beginning of a start bit. The "quiet" time between transmissions lets it determine when to "look" for the transition at the edge of a start bit. Note that a standard UART follows the start bit with the least-significant bit (LSB) in the byte transmitted. So, the transmission of the ASCII value for the character "J" (0x4A or 01001010_2) looks like the bits shown in Figure 16.3.

Software for the Arduino Uno and ARM mbed modules make it easy to set up a UART and serial communications. In its factory-default condition, an XBee communicates through its internal UART at 9600 bits/second with eight data bits, no parity, and one stop bit. The experiments in this book use these settings, but UARTs on MCUs and XBee modules can communicate at higher rates.

Important: When you run into a communication problem that involves UARTs, always check the serial-port configurations first. Many problems involve mismatched settings you assume are properly set, but aren't. In particular, in terminal-emulator software ensure you have no flow control.

TERMINAL-EMULATION SOFTWARE

I used the Windows HyperTerminal program in the following steps. You can find other terminal-emulation programs for PCs on the Internet. Before you use a terminal-emulator program, quit X-CTU to avoid contention for use of the same PC COM port that connects to an XBee module at your PC.

Set your terminal emulator for a data rate of 9600 bits/second, or 9600 baud. Set the number of stop bits to 1, parity to none, and number of data bits to 8. If you can select a flow-control mode, choose none. Ensure you have the terminal emulator set for the COM port that connects to the XBee RCVR module. If you cannot determine the correct COM port assignments, go to Windows Settings→Control Panel→System→Hardware→Device Manager→Ports (COM & LPT) to identify available COM devices. Then try the available COM ports in your terminal-emulator software.

A terminal-emulator program that receives characters from an external device will continue to print them on one line in its window unless instructed otherwise. Your programs must send the terminal emulator a "line feed" command, also called "new line," to move the cursor down a line in the terminal window, followed by a "carriage return" command to move the cursor to the left side of the window. You might see carriage-return and line-feed abbreviated as CR and LF or together as CRLF.

Use the hex value 0x0A to cause a line feed and 0x0D for a carriage return. You can send these bytes directly or in a C-language program use the "\n" and "\r" symbols in `print` commands. Depending on your terminal-emulator settings, you might need to try 0x0A and 0x0D together or separately.

Important: If you plan to go to Experiment 17 now, you can leave your hardware set up when you finish this experiment.

USING AN ARDUINO UNO

Uno Step 1. The Arduino Uno module uses code written in the C language to communicate with an external device via a UART. A group of `Serial` commands such as `Serial.print` and `Serial.begin` simplify serial-communication code. The Arduino Uno module always assumes an 8-bit transmission with no parity and one stop-bit, abbreviated as 8N1. A software command sets the bit rate independently.

The Arduino Uno module defines pin 1, labeled TX-- > 1 on the board, as its UART transmitter output, so this experiment uses that pin. Pin numbering on a Uno module starts with pin 0. *Do not count pins, use the pin labels and numbers printed on the Uno board.*

The C program EX16_Uno_Hello will send out—via the UART—the ASCII characters H, e, l, l, o, along with three periods and two spaces. Do not enter this program yet.

Uno Program EX16_Uno_Hello

```
void setup()
{
   Serial.begin(9600);              //Set up serial port
}

void loop()
{
   Serial.print("Hello... ");
   delay(100);                      // wait for 0.1 second
}
```

The `Serial.begin` command set the data rate at 9600 bits/sec. The rest of the program simply goes through an infinite loop that continues to display "Hello..." again and again in the terminal-emulator window. The `delay` command provides a 100-millisecond (0.1-second) delay so you can see the message arrive on the screen.

Uno Step 2. The diagram in Figure 16.4 shows the Arduino Uno connections to your XBee XMTR module. Note the use of the 5-to-3 logic-level-conversion circuit (LLC).

The logic-level conversion circuit described in Appendix A requires additional connections not shown in the Figure 16.4 diagram. Please make the following additional connections if they do not yet exist at your logic-level-conversion circuit. On the 5-to-3-volt logic-level conversion circuit,

- Connect a 1000-ohm resistor (brown-black-red) between pins 1 and 2.
- Connect pins 11, 12, 13, and 22 to ground.
- Connect pin 1 to +5 volts.
- Connect pins 23 and 24 to +3.3 volts.

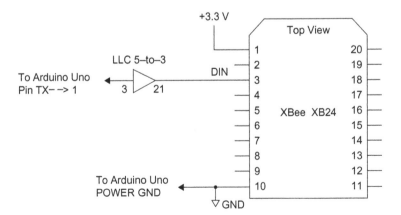

FIGURE 16.4 Schematic diagram for data transmission from an Arduino Uno module to an XBee transmitter.

Important: Make these extra connections *only* at the 5-to-3-volt logic-level-conversion circuit. Do not make any of these connections to the XBee module or you will damage it.

People often run into problems because their circuits lack a common ground. You must ensure you have a ground connection that links all power-supply and circuit grounds used in an experiment. In this case, you need a common ground between the Arduino Uno module, the logic-level-conversion circuit, the XBee breadboard, and your power supplies. You do not need a common ground between your breadboarded circuits and your PC because they have no common electrical connection. All communications occur over the wireless link.

Turn on power to your Arduino Uno module and temporarily remove the end of the wire at the TX-->1 pin on the arduino module. Breaking this connection prevents problems due to the Arduino Uno module sharing its serial port with the USB connection used to download programs from your PC.

Load into the arduino uno the EX16_Uno_Hello program shown earlier. Turn on power to your logic-level-conversion circuit and to the XBee module in the breadboard. Reconnect the wire from your logic-level-conversion circuit to the Arduino Uno module TX receptacle. Whenever you load programs into an Arduino Uno module, you must disconnect the TX and RX connections, and reconnect them after the program has loaded.

In your terminal emulator you should see the repeating pattern: Hello... Hello... Hello... This display indicates the Uno module created the serial bytes that represent each character in the message. The XMTR module connected to the Uno module transmitted this information via its wireless output and the RCVR module received it. The RCVR module sent the information to the PC's terminal emulator via its USB connection.

Keep in mind the terminal emulator must communicate with the virtual serial port that connects to the XBee RCVR module and *not* to the virtual serial port that connects from your PC to the Arduino Uno board.

If you did not see the repeated "Hello..." message, you likely had incorrect UART settings for the Arduino Uno MCU or in your terminal emulator. Recheck your settings so they match 9600 bits/second, or 9600 baud. You also need eight data bits, no parity, and one stop bit. Set the terminal-emulator software for no flow control.

Uno Step 3. In this step you will change the program slightly so the MCU creates its own bytes and the terminal emulator will display many different characters. As you program the Uno module, you can leave power applied to your circuits. Remove the TX connection at the Arduino Uno receptacle strip. Load the new program, EX16_Uno_Alpha, into your Uno module and run it. Reconnect the wire to the TX receptacle.

At times when using an Arduino Uno module I had difficulty getting the Arduino software to recognize the board and I also saw error messages such as "Serial port 'COMxx' already in use." If you see error messages, close all other programs such as X-CTU or the terminal emulator that might try to connect with the virtual serial port used to connect the Arduino compiler to the Uno board. It might take several attempts, as I described in the Introduction, to get an Arduino Uno module to respond properly.

Uno Program EX16_Uno_Alpha

```
char alphachar;

// initialize serial port
void setup()
{
   Serial.begin(9600);
}

void loop()
{
   alphachar = 0x20;
   while (alphachar < 0x7F)
   {
      Serial.print(alphachar,BYTE);
      delay(100);
      alphachar = alphachar + 1;
   }
}
```

What do you see in the terminal-emulator window?

This program sets up a loop to increment a value that starts at 0x20. The `while` loop tests `alphachar` to determine if it is less than 0x7F. When `alphachar` exceeds 0x7E, the program exits the `while` loop and the main `loop` starts again and initializes `alphachar` back to 0x20.

The values correspond to the characters "space" through "~" in the ASCII character set, and after "printing" one set, the code goes through the same sequence again and again in an endless, or infinite, `loop` that runs until you turn off power.

Uno Step 4. Can you write a short C program to print only the numbers from 9 to 0 and then start over at 9 and go down to 0 again and again? Each sequence, 9876543210, should appear on a separate line in a terminal-emulator window. You can use "\r\n" in a `Serial.print` command to force a new line in your terminal emulator. The downloaded code file for this experiment includes a program that prints numbers in this format.

If you plan to go right to Experiment 17, you may leave your circuits set up. You will use them as set up for this experiment.

USING AN ARM MBED MODULE

mbed Step 1. The ARM mbed (yes, all lowercase letters) module uses code written in the C language to communicate with an external device via its UART. You can give the serial port a name, select the pins used to transmit and receive information, establish the UART operating conditions, and then write code with C commands such as `print` and `printf`. The ARM mbed module's serial ports default to the format that uses eight data bits, no parity, and one stop bit. Refer to the ARM mbed online "Handbook" for information about all serial-port commands: www.mbed.org.

The C program EX16_mbed_Hello will send via the ARM mbed UART the ASCII characters H, e, l, l, o, along with three periods and two spaces. The ARM mbed module offers three serial ports and the program below uses pin 9 to transmit data. You won't use the receiver input, pin 10, in this experiment. The pin numbers refer to the module and not to the on-board MCU. Don't enter this program yet.

mbed Program EX16_mbed_Hello

```
#include "mbed.h"

Serial XBeePort(p9, p10);   // set transmit & receive pins

int main()
{
   XBeePort.baud(9600);     //Set bit rate

   while(1)
   {
      XBeePort.printf("Hello... ");
      wait_ms(100);
   }
}
```

The line # include "mbed.h" causes the ARM mbed compiler to use a file that defines many operations so you do not have to set or clear bits in control registers to set up a UART or serial port.

The Serial XBeePort(p9, p10); command names the serial port as XBeePort to make it easy to identify and refer to in code. You could name the serial port almost anything, such as framostan, SerialX, MySerialPort, and so on. After naming the serial port you use its name in the main routine to set its bit rate.

The XBeePort.printf("Hello..."); command sends the word Hello, followed by three periods and two spaces to your terminal-emulator window. The wait_ms(100); command causes a 100 millisecond (0.1 second) delay.

mbed Step 2. The diagram in Figure 16.5 shows the ARM mbed module connections to your XBee XMTR module. Because an ARM mbed module provides 3.3-volt logic levels, you do not need to convert voltage levels and can make direct connections between an ARM mbed and an XBee module. The ARM mbed module obtains power from your PC via its USB connection.

Connect your ARM mbed module to your PC and load the program EX16_ mbed_Hello into it. Note: An ARM mbed module acts like an external USB drive, so you simply "save" your code to the ARM mbed module and then press the ARM mbed pushbutton to start your program. On my PC, the ARM mbed module appeared as the F: drive. Refer to the mbed.org web site for tutorial and reference information.

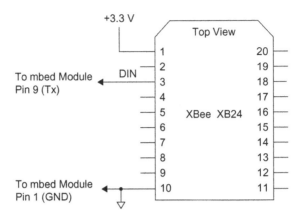

FIGURE 16.5 The ARM mbed schematic diagram for Experiment 16.

After you turn on power to your XBee module in the breadboard, press the Reset button on the ARM mbed to start the program you just downloaded. In your terminal emulator you should see the repeating pattern: Hello... Hello... Hello... and so on. This display indicates the ARM mbed module created the serial bytes that represent each character in the message. The XMTR module connected to the ARM mbed module transmitted this information via its wireless output and the RCVR module received it. The RCVR module sent the information to the PC's terminal emulator.

Important: People often run into problems because their circuits lack a common ground. You must ensure you have a ground connection between the ARM mbed module and the XBee breadboard.

mbed Troubleshooting

If you did not see the repeated "Hello..." message, you likely had incorrect UART settings in the ARM mbed MCU or in your terminal emulator. Recheck your settings so they match 9600 bits/second, or 9600 baud. You also need eight data bits, no parity, one stop bit, and no flow control in your terminal-emulator program.

Because the ARM mbed uses a USB port to communicate with its Web-based development software, ensure you have selected the "USB Serial Port" for the USB-to-XBee adapter board when you set up your terminal emulator. Do not use the COM port for the ARM mbed module. That COM port serves as the programming and debugging port for the ARM mbed compiler. If you cannot determine the correct COM port assignments, go to Windows Settings→ Control Panel→ System→ Hardware→ Device Manager→ Ports (COM & LPT) to identify available COM devices.

mbed Step 3. In this step you will change the program slightly so the MCU creates its own bytes within the program. The terminal emulator will display many different characters.

As you program the ARM mbed module, you can leave power applied to your circuits. You will not change any settings in your XMTR or RCVR just because you have power on. Load program EX16_mbed_Alpha into your ARM mbed module and run it.

mbed Program EX16_mbed_Alpha

```
#include "mbed.h"

Serial XBeePort(p9, p10);   // select UART pins

int alphachar;

int main()
{
    XBeePort.baud(9600);

    while(1)
    {
        XBeePort.printf("\n\r");
        alphachar = 0x20;
            while(alphachar <0x7F)
            {
                XBeePort.putc(alphachar);
                alphachar = alphachar + 1;
                wait_ms(100);
            }
    }
}
```

What do you see in the terminal-emulator window?

This program sets up a loop that will increment the variable alphachar that starts with a value of 0x20 and ends with with a value of 0x7E. The while loop tests alphachar to determine if it is less than 0x7F. When alphachar exceeds 0x7E, the program exits the while(alphachar... loop and the while(1) loop starts again and initializes alphachar back to 0x20.

The values correspond to the characters "space" through "~" and after printing one set of characters, the ARM mbed code goes through the same sequence again and again in an endless, or infinite, loop that runs until you turn off power. Note a difference between the "Hello" program and the program directly above. The EX16_mbed_Alpha program uses the XBeePort. putc command to send a single hex byte—the alphachar value—to the UART. This program used an XBeePort.printf command to send the terminal emulator commands for a carriage return and a line feed.

mbed Step 4. Can you write a short C program to print only the numbers from 9 to 0 and then start over at 9 and go to 0 again and again? Each sequence, 9876543210, and so on should appear on a separate line in the terminal-emulator window. The downloaded code file for this experiment includes a program that prints numbers in this format.

If you plan to go right to Experiment 17, leave your circuits set up. You will use them as set up for this experiment.

Two-Way Communications with XBee Modules

REQUIREMENTS

2 XBee modules
1 XBee adapter board
1 Solderless breadboard
1 USB-to-XBee adapter
1 USB cable (type depends on MCU board chosen)
1 Microcontroller board with a serial port (see Experiment 16)
1 5-V-to-3.3-V logic-conversion circuit or module (see text)
1 3.3-V-to-5-V logic-conversion circuit or module (see text)
1 1000-ohm, 1/4-watt, 10% resistor (brown-black-red) (see text)
Terminal-emulation software such as HyperTerminal for Windows
Digi X-CTU software running on a Windows PC, with two open USB ports

INTRODUCTION

In Experiment 16 you learned how a simple computer program could use a remote XBee module to transmit information to a receiving XBee module. In this experiment you will use two-way communications to send a wireless message from your PC to a remote microcontroller (MCU), after which the MCU will send back a reply. XBee modules operate most of the time as receivers, but they can quickly change to the transmit mode and then back to the receive mode under control of an MCU.

The diagram in Figure 17.1 shows the module configuration for this experiment. Note the two-way communications between the XMTR and RCVR XBee modules and between the XMTR XBee module and the MCU. This diagram represents a typical configuration for XBee modules, as well as for other types of transceivers.

The Hands-on XBee Lab Manual.

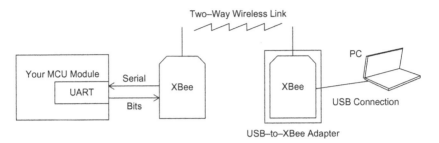

FIGURE 17.1 A block diagram of computer-to-computer communications.

This experiment lets you work with either of two MCU modules: An Arduino Uno MCU module or an ARM mbed MCU module. These modules and boards are available at low cost and provide easy ways to run experiments without getting deeply into hardware or software details.

If you have just completed Experiment 16, please jump ahead to Step 3 below. If you have not yet run Experiment 16, I recommend you do so before you run this experiment.

Step 1. In this step you will configure the XMTR module to operate with its default factory settings. Ensure you have turned off power to the breadboard. Then remove the XMTR module from the XBee socket in your breadboard and place it in the USB-to-XBee adapter and reconnect this adapter to the USB cable and your PC.

If not already running, start the X-CTU software. Within the X-CTU window, click on Modem Configuration and then Restore. After a few seconds you should see the message "Restore Defaults..complete" appear. Click on Read to obtain the restored default values from the XMTR module. Check to ensure the DL – Destination Address Low and MY – 16-Bit Source Address each have a value of 0. This condition indicates the X-CTU software has reset all XMTR module settings to their default state. If DL and MY are not equal to 0, retry the procedure in this step. After you confirm the default condition, place the XMTR module back in its adapter on the solderless breadboard.

Step 2. Place the RCVR module in the USB-to-XBee adapter and reconnect this adapter to the USB cable. Perform the configuration described in the second paragraph of Step 1. After you have restored the default settings for the RCVR XBee module, leave it in the USB-to-XBee adapter connected to your PC.

Step 3. The schematic diagram in Figure 17.2 shows the connections needed between an XBee module and an MCU for serial communications. Later you will make specific connections for the MCU you chose to work with. For now, if you have other connections to the XMTR module, please disconnect them.

SOFTWARE

This experiment uses two programs with the Arduino Uno or ARM mbed MCU modules. The first program sets up the MCU and the XBee module (XMTR) to accept five bytes sent from the XBee module (RCVR) attached to your PC.

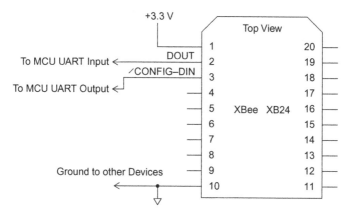

FIGURE 17.2 Connections at an XBee XMTR module for 2-way MCU serial communications.

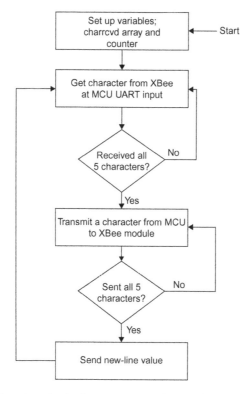

FIGURE 17.3 Flow chart for the first program used to receive and retransmit five bytes of information.

As soon as the MCU module has received all five bytes, it transmits them back to your PC. You will use a terminal-emulator program or the Terminal window in the X-CTU software to enter characters, each of which exists as a byte of information. The flow chart in Figure 17.3 shows how the program operates.

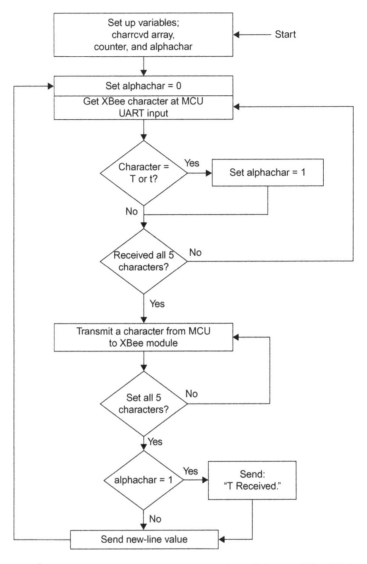

FIGURE 17.4 Flow chart for the extended program that will detect a "T" or "t" character received by the XMTR XBee module.

In the first program, the MCU simply receives five characters and transmits them back to the terminal emulator. The second program, shown as a flow chart in Figure 17.4, makes a decision based on the bytes received by the XMTR XBee module. The MCU will still accept the bytes for five typed characters and retransmit them. But if you type a "T" or a "t," in the group of five characters, the reply will include the message "T Received."

This type program demonstrates how an MCU can take an action based on the information it receives. If an MCU receives a "T," it might make a temperature measurement, convert it to degrees Celsius, and transmit that value back to the requesting XBee module. Likewise, receipt of a "P" might trigger a response with atmospheric pressure from a remote XBee-and-MCU location.

Experiments that use an MCU board will provide complete listings, so you will not need to write programs. Download all the code from: http://www .elsevierdirect.com/companion.jsp?ISBN=9780123914040.

TERMINAL-EMULATION SOFTWARE

I used the Windows HyperTerminal program in the following steps. You can find other terminal-emulation programs for PCs on the Internet. Before you use a terminal-emulator program, quit X-CTU to avoid contention for use of the same PC COM port that connects to an XBee module at your PC.

Set your terminal emulator for a data rate of 9600 bits/second, or 9600 baud. Set the number of stop bits to 1, parity to none, and number of data bits to 8. If you can select a flow-control mode, choose none. Ensure you have the terminal emulator set for the COM port that connects to the XBee RCVR module. If you cannot determine the correct COM port assignments, go to Windows Settings→ Control Panel→ System→ Hardware→ Device Manager→ Ports (COM & LPT) to identify available COM devices. Then try the available COM ports in your terminal-emulator software.

Depending on your terminal-emulator setting, you might not see typed characters in its window. To set HyperTerminal to "echo" what you type, open it and follow these menu choices: File→Properties→Settings→ASCII Setup, and click on "Echo typed characters locally." Other emulators might have a similar capability.

USING AN ARDUINO UNO MODULE

Arduino Uno Step 1. The Arduino Uno module uses code written in the C language to communicate with an external device via a UART, and commands such as `Serial.print` and `Serial.begin` simplify serial-communication software. The Arduino Uno module always assumes an 8-bit transmission with no parity and one stop-bit, abbreviated as 8N1.

The C-language program EX17_Uno_Five will monitor the Arduino Uno UART input for a new byte received from the XMTR XBee module. When the MCU UART receives a byte, it will save it in an array that can hold five values. After the MCU has received five bytes, it will send them back to the XMTR module that will transmit them wirelessly to the RCVR module. The RCVR module will send the bytes to your PC, which will display them in its terminal-emulator window. Do not enter program EX17_Uno_Five at this time.

Arduino Uno Program EX17_Uno_Five

```
/*
 Experiment 17 Arduino Uno
 EX17_Uno_Five.pde
 Jon Titus, 07-06-2011
 */

char charrcvd[5];                       // Set up array for rcvd bytes
int counter;                            // Set counter variable

void setup()                            // Set serial port for 9600
                                        // bits/sec
{
  Serial.begin(9600);
}

void loop()                             // main loop starts here
{
  counter = 0;                          // set counter to zero to start
  while(counter < 5)                    // while count is less than 5
  {                                     // go through this loop
    if (Serial.available() > 0)         // check UART for any new data
    {
      charrcvd[counter] = Serial.read();  // detected new UART
                                        // data, so get
                                        // the UART byte
      counter = counter + 1;            // increment counter by one
    }
  }                                     // received five characters,
                                        // so exit while loop

  counter = 0;                          // set counter to zero to start
  while(counter < 5)                    // while counter is less than 5
  {                                     // go through this loop
    Serial.print(charrcvd[counter],BYTE);    // print character
                                        // as a byte code
    delay(100);                         // wait 100 msec between
                                        // characters
    counter = counter + 1;              // increment counter by one
  }
Serial.print("\n\r");                   // "print" new line
}                                       // printed five characters,
                                        // so exit while loop

//end of code, but execution returns to loop()
```

The Serial.begin command sets the Arduino Uno UART data rate at 9600 bits/second. The first while(counter < 5) loop will save five bytes received from the UART and this loop includes the statement:

```
if (Serial.available() > 0)
```

The program needs this statement to determine whether or not the UART has a *new* byte available. The Serial.available() function returns the value 0 if the UART receiver is empty and it returns a non-zero value when the

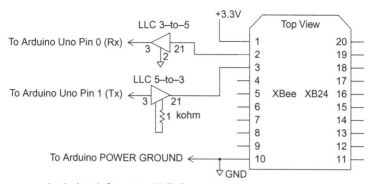

Logic–Level–Converter (LLC) Connections:
Ground at pins 11, 12, 13, 22.
+5 volts at pin 1; +3.3 volts at pins 23, 24.
On LLC 5–to–3: 1 kohm resistor between pins 1 and 2.
On LLC 3–to–5: ground pin 2.

FIGURE 17.5 Arduino Uno schematic diagram with logic-level-converters.

UART has a new byte available. Without this statement, the software would continuously read the UART receiver, regardless of whether or not it contained new data, and cause the program to display whatever byte the UART receiver holds. An Arduino Uno program MUST check the UART for available data before it tries to read data from it. Other MCUs behave in a similar fashion; that is, they provide a "flag" that software can test to determine if a UART or other device is ready or not ready, full or empty, on or off, and so on.

The program EX17_Uno_Five will save five bytes in the `charrcvd` array and then transmit them back to the XMTR module so they appear via wireless link in your terminal-emulator window. The `delay` command provides a 100-millisecond (0.1-second) delay so you can see the five characters arrive on the screen one by one.

Arduino Uno Step 2. The schematic diagram in Figure 17.5 shows the connections from an Arduino Uno module to logic-level-converter circuits, and to an XBee module. Note the use of separate logic-conversion circuit for 5-V-to-3-V logic and for 3-V-to-5-V logic. Pay attention to the notes below and in Figure 17.5 that explain additional connections not shown in the diagram for the sake of clarity:

- On LLC 5-to-3, connect a 1-kohm resistor (brown-black-red) between pins 1 and 2.
- On LLC 3-to-5, connect pin 2 to ground.
- On both logic-level-converter devices, connect pins 11, 12, 13, and 22 to ground.
- On both logic-level-converter devices, connect pin 1 to +5 volts.
- On both logic-level-converter devices, connect pins 23 and 24 to +3.3 volts.

Note that pin numbers on an Arduino Uno board start with pin 0. DO NOT count pins, use the pin designation and numbers printed on the Arduino Uno board.

Important: People often run into problems because their circuits lack a common ground. You must ensure you have a ground connection that links all the power-supply and circuit grounds used in an experiment. In this case, you need a common ground between the Arduino Uno module, the logic-translation circuits, and the XBee breadboard.

Arduino Uno Step 3. Turn on power to your breadboard and to the logic-level-converter circuits.

Label the wire, or use a colored wire, from the logic-level-converter circuit to the Arduino Uno pin 0 (RX) and remove it from the Arduino Uno connector. Leave the other end connected to the logic-level-converter circuits. Likewise, label the wire from the logic-level-converter circuit to the Arduino Uno pin 1 (TX) and remove it from the Arduino Uno connector. Leave the other end connected to the logic-level-converter circuits. You will reconnect these wires after you download a program to the Arduino Uno MCU. (The USB connection also uses the MCU UART, so temporarily removing the two wires eliminates the possibility of conflicts caused by simultaneous use of the UART pins by the USB connection and the logic-level-converter circuits.)

Load the program EX17_Uno_Five into the Arduino compiler, compile it and load it into the MCU.

Replace the wires in the Arduino Uno connector at positions Arduino Uno pin 0 (RX) and Arduino Uno pin 1 (TX). Press the Arduino Uno Reset pushbutton.

Arduino Uno Step 4. In your terminal emulator you should see a blank text window. (If you just completed Experiment 16, you might see information remaining from old transmissions.)

Type four characters, such as ASDF, into your terminal emulator. Depending on your emulator setting, you might not see these characters in its window. Now type a fifth character. You should see the characters appear in the emulator's window. (Some terminal emulator programs do not "echo" characters as you type them. So you might not see characters as you type them. Only the characters sent from the MCU would appear at your PC. HyperTerminal does not "echo" what you type, but the X-CTU Terminal window does.)

The Arduino Uno module received the five characters and saved them in an array. Upon receipt of the fifth character, the MCU transmitted the five characters back to the terminal emulator via the XMTR to the RCVR module.

Arduino Uno Troubleshooting. If you did not see the characters, you likely had incorrect UART settings in the Arduino Uno MCU or in your terminal emulator. Recheck your settings so they match 9600 bits/second or 9600 baud. You also need eight data bits, no parity, and one stop bit, and the Arduino Uno will default to these settings. Also ensure you have a common ground connection among all your circuits.

Arduino Uno Step 5. In the steps above, the Arduino Uno module simply received five characters and transmitted them back to the terminal emulator. But the MCU also could interpret received information and take actions based on commands you send it. The next program, EX17_Uno_FiveT, modifies the earlier Arduino Uno code so the MCU will recognize receipt of a "T" or a "t" (without the quotes). The MCU will respond with the five characters, and will transmit a short message to acknowledge it received "T" or "t" among the characters.

Arduino Uno Program EX17_Uno_FiveT

```
/*
 Experiment 17 Arduino Uno
 EX17_Uno_FiveT.pde
 Jon Titus, 07-07-2011
*/

char charrcvd[5];                   // Set up array for rcvd bytes
int counter;                        // Set counter variable
char alphachar;                     // Use a byte for a flag value

void setup()                        // Set serial port for 9600
                                    // bits/sec
{
  Serial.begin(9600);
}

void loop()                         // main loop starts here
{
  alphachar = 0;                    // clear the alphachar flag
  counter = 0;                      // set counter to zero to start
  while(counter < 5)                // while count is less than 5
  {                                 // go through this loop
    if (Serial.available() > 0)     // check UART for any new data
    {
      charrcvd[counter] = Serial.read();  // detected new UART
                                    // data, so get
                                    // URAT byte
      if ((charrcvd[counter] == 0x54) || (charrcvd[counter] ==
                                    // 0x74))
                                    //test for T or t
      {
        alphachar = 1;             // found T or t, so set
                                    // alphachar flag
      }
      counter = counter + 1;        // increment counter by one
    }
  }                                 // received five characters,
                                    // so exit while loop

  counter = 0;                      // set counter to zero to start
  while(counter < 5)                // while counter is less than 5
  {                                 // go through this loop
    Serial.print(charrcvd[counter],BYTE);    // print character
                                    // as a byte code
    delay(100);                     // wait 100 msec between
                                    // characters
    counter = counter + 1;          // increment counter by one
  }
Serial.print("\n\r");               // "print" new line
  if (alphachar == 1)               // if alphachar flag set,
                                    // print message
  {
    Serial.print("T Received.\n\r");
  }
}                                   // printed five characters,
                                    // so exit while loop

//end of code, but execution returns to loop()
```

The added steps first set the variable alphachar to 0 and check received characters to determine if they match the hex value for "T" or "t." If a character code matches either 0x54 or 0x74, alphachar gets set to 1. Thus alphachar serves as a "flag" that lets the MCU know whether or not the received characters include "T" or "t." Note the double upright bars, ||, are not the characters for the numeral 1 or lowercase Ls. Find the "bar" character on most keyboards with the backslash character, \. The two bars indicate a logical OR condition within the if statement. So, if either the (charrcvd[counter] == 0x54) OR the (charrcvd[counter] == 0x74) condition is true, the program assigns alphachar the value 1.

After the program transmits the five characters back to the XMTR module, and thus to your terminal emulator, it uses an if statement; if (alphachar == 1) to test alphachar. If alphachar equals 1, which means the five characters contained a "T" or a "t," the Arduino Uno transmits the message "T Received." Note in the C language, the double equal-sign notation = = indicates a test for equality between two values. If you want to assign a value to a variable, use = and not ==.

Temporarily remove the wires inserted into Arduino Uno pin 0 (RX) and into Arduino Uno pin 1 (TX). You will reconnect these wires after you download a program to the Arduino Uno MCU.

Load the program EX17_Uno_FiveT into the Arduino compiler, compile it and load it into the MCU.

Replace the wires in the Arduino Uno connector at positions Arduino Uno pin 0 (RX) and Arduino Uno pin 1 (TX). Press the Arduino Uno Reset pushbutton.

When you run the EX17_Uno_FiveT program, you should see the same results obtained with the first program except that typing a "T" or a "t" will cause the reply to include the "T Received." message.

Please jump ahead to "Control of Remote XBee Modules with MCU AT Commands" at the end of the ARM mbed section of this experiment.

USING AN ARM MBED MODULE

ARM mbed Step 1. The ARM mbed module uses code written in the C language to communicate with an external device via its UART. You can give the serial port a name, select the pins used to transmit and receive information, establish the UART operating conditions, and then write code with C commands such as print and printf. The ARM mbed module's serial ports default to the format that uses eight data bits, no parity, and one stop bit. Refer to the ARM mbed online "Handbook" for information about all serial-port commands: www.mbed.org.

The C-language software for program EX17_mbed_Five will monitor the MCU's UART to "watch" for information received from the XMTR module. When the MCU UART signals the receipt of a byte, it will save it in an array. After the MCU receives five bytes, it will send them to the XMTR module that will transmit them wirelessly to the RCVR module. The RCVR module will send the bytes to your PC, which will display them in its terminal-emulator window. Don't enter program EX17_mbed_Five at this time.

ARM mbed Program EX17_mbed_Five.cpp

```cpp
/*
Experiment 17 ARM mbed
EX17_mbed_Five
Jon Titus, 07-07-2011
*/

#include "mbed.h"

Serial XBeePort(p9, p10);              // assign UART pins
                                       // 9 and 10
int charrcvd[5];                       // set up array for rcvd
                                       // bytes
int counter;                           // set counter variable

int main()                             // main program starts
                                       // here
{
    XBeePort.baud(9600);               // set UART bit rate

    while(1)                           // outside loop starts
                                       // here
    {
        counter = 0;                   // set counter to 0
        while(counter <5)              // while count less than 5
        {                              // go through this loop
            if (XBeePort.readable())   // does UART have a byte?
            {
                charrcvd[counter] = XBeePort.getc();  //yes, get
                                                      // UART byte
                counter = counter + 1;                // increment
                                                      // counter
            }
        }
        XBeePort.printf("\n\r");        // rcvd 5 bytes, so "type"
                                       // new line
        counter = 0;                   // reset counter to 0
        while(counter < 5)             // while count less than 5
        {                              // go through this loop
            XBeePort.putc(charrcvd[counter]);         //print
                                       // character as a byte
            counter = counter + 1;     // increment counter
            wait_ms(100);
        }
        XBeePort.printf("\n\r");        // printer all 5
                                       // characters, "type"
    }                                  // a new line
}
//end of code but execution returns to outside loop
```

The line #include "mbed.h" causes the ARM mbed compiler to use a file that defines many operations so you do not have to control individual MCU registers to configure the UART.

The while(1) loop will run "forever," or until you change your code or turn off the ARM mbed module. Within the while(1) loop, the first while(counter <5) loop uses the command XBeePort.readable() to determine when the UART has received a character. You don't want to try to get data from the UART if it contains nothing new. Other MCUs behave in a similar fashion; that is, they provide a "flag" that software can test to determine if a UART or other device is ready or not ready, full or empty, on or off, and so on.

When the program determines the UART has a new byte it stores that byte in the charrcvd array until the array contains five values.

Next, the last while(counter <5) loop gets the bytes stored in the charrcvd array and sends them back to the XMTR XBee module one at a time. The module immediately transmits the wireless information to the RCVR module. The wait_ms(100) command causes a 100-millisecond (0.1-second) delay between transmitted characters, so they appear one at a time in the terminal-emulator window as transmitted from the XMTR to the RCVR module.

ARM mbed Step 2. The circuit diagram Figure 17.6 shows the ARM mbed module connections to the XBee XMTR module. Make these connections now. Because an ARM mbed module provides 3.3-volt logic levels, you do not need logic-level-converter circuits. The ARM mbed module obtains power from its USB connection.

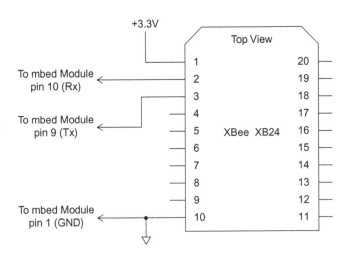

FIGURE 17.6 ARM mbed-to-XBee circuit connections.

Important: People often run into problems because their circuits lack a common ground. You must ensure you have a ground connection between the ARM mbed module and the XBee breadboard.

Connect your ARM mbed module to your PC via a USB cable and load the program EX17_mbed_Five.cpp into the compiler. Compile the program and load it into the MCU. Note: An ARM mbed module acts like an external USB drive, so you simply "save" your code to the ARM mbed module. On my PC, the ARM mbed module appeared as the F: drive. Refer to the mbed.org web site for tutorial and reference information.

Turn on power to your XBee RCVR module in the breadboard. Push the ARM mbed pushbutton to start your program. In your terminal emulator you should see a blank text window. (If you just completed Experiment 16, you might see information remaining from old transmissions.) Type four characters, such as ASDF, into your terminal emulator. Now type a fifth character. You should see all five characters appear in the emulator's window.

The ARM mbed module received the five characters and saved them in an array. Upon receipt of the fifth character, the ARM mbed transmitted the five characters back to the terminal emulator via the XMTR and RCVR modules.

ARM mbed Troubleshooting: If you did not see the five-character response from the ARM mbed module, you likely had incorrect UART settings in the ARM mbed MCU or in your terminal emulator. Recheck your settings so they match 9600 bits/second, or 9600 baud. You also need eight data bits, no parity, and one stop bit, along with no flow control.

Because the ARM mbed uses a USB port to communicate with its Web-based development software, ensure you have selected the "USB Serial Port" for the USB-to-XBee adapter board when you set up your terminal emulator. Do not use the COM port for the ARM mbed module. That COM port serves as the programming and debugging port for the ARM mbed compiler.

Recheck your circuit to ensure you have a common ground between all circuit elements. Also check the transmit and receive wires to ensure they connect to the proper points.

ARM mbed Step 3. In the steps above, the ARM mbed module simply received five characters and transmitted them back to the terminal emulator. But, the ARM mbed also could interpret received information and take actions based on the commands it receives. The next program modifies the EX17_mbed_Five program to recognize receipt of a "T" or a "t." The MCU will still transmit the five characters it received, and will transmit a short message to acknowledge it received "T" or "t" among the characters.

ARM mbed Program EX17_mbed_FiveT.cpp

```
/*
Experiment 17 ARM mbed
EX17_mbed_FiveT
Jon Titus, 07-07-2011
*/

#include "mbed.h"

Serial XBeePort(p9, p10);              // assign UART pins
                                       // 9 and 10
int charrcvd[5];                       // set up array for
                                       // rcvd bytes
int counter;                           // set counter variable
int alphachar;                         // set up flag value
int main()                             // main program starts
                                       // here
{
    XBeePort.baud(9600);               // set UART bit rate

    while(1)                           // outside loop starts
                                       // here
    {
        alphachar = 0;                 // set flag value to 0
        counter = 0;                   // set counter to 0
        while(counter <5)              // while count less than 5
        {                              // go through this loop
            if (XBeePort.readable())   // does UART have a byte?
            {
                charrcvd[counter] = XBeePort.getc();  //yes, get
                                                      //UART byte
                if ((charrcvd[counter] == 0x54) || (charrcvd
                                    // [counter] == 0x74))
                    {
                        alphachar = 1;
                    }
                counter = counter + 1;              // increment
                                                    // counter
            }
        }
        XBeePort.printf("\n\r");        // rcvd 5 bytes, so
                                        // "type" new line
        counter = 0;                    // reset counter to 0
        while(counter < 5)              // while count less than 5
        {                               // go through this loop
            XBeePort.putc(charrcvd[counter]);       //print
                                        // character as a byte
            counter = counter + 1;      // increment counter
            wait_ms(100);
        }
        XBeePort.printf("\n\r");        // printer all 5
                                        // characters, "type"
                                        // new line
```

```
if (alphachar == 1)              // if alphachar = 1,
       {                          // print "T Received." message
          XBeePort.printf("T Received.\n\r");
       }
   }
}
//end of code but execution returns to outside loop
```

The added steps in the new program, EX17_mbed_FiveT, first set the variable `alphachar` to 0. Instructions then check received characters to determine if they match the hex value for "T" or "t." If a character code matches either 0x54 or 0x74, `alphachar` gets set to 1. Thus, `alphachar` serves as a "flag" that lets the computer know whether or not the received characters include a "T" or a "t." Note the double upright bars, ||, are not the characters for the numeral 1 or lowercase Ls. Find the "bar" character on most keyboards with the backslash "\" character. The two bars indicate a logical OR condition within the `if` statement. So, if either the `(charrcvd[counter] == 0x54)` OR the `(charrcvd[counter] == 0x74)` condition is true, the program assigns `alphachar` the value 1.

After the program transmits the five received characters back to the XMTR module, and thus to your terminal emulator, it uses an `if` statement; `if (alphachar == 1)`. If `alphachar` equals 1, which means the five characters contained a "T" or a "t," the MCU transmits the message, "T Received." Note that in the C language, the notation = = indicates a test for equality between two values. If you want to assign a value to a variable, use = and not = =.

CONTROL OF REMOTE XBEE MODULES WITH MCU AT COMMANDS

In this experiment you saw how an MCU connected to a remote XBee module could respond to a command represented by the letter T or t. Could you have the MCU control the remote XBee module to turn on an LED at the AD1-DIO1 pin when it has received the letter T or t? Assume the LED connects from +3.3 volts through a resistor to pin 19 (Figure 17.7).

When you send an API command packet to change the AD1-DIO1 pin to a logic-0 condition, you use the command D1 4, or in hex 0x44 0x34 0x04. So you might think an API command would work in this situation, too. Unfortunately, configuration instructions for the remote XBee module in this experiment did not include a step to enable the API capability. But even though the XBee module operates in transparent mode with the factory-default values, it will still respond to AT commands just as if sent from the X-CTU Terminal window. The following example shows the ATD1 command used to change the AD1-DIO1 pin to a logic 0 on a module attached to a PC:

```
+++OK       2B 2B 2B 4F 4B 0D

ATD14       41 54 44 31 34 0D

OK          4F 4B 0D
```

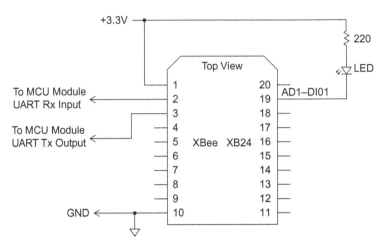

FIGURE 17.7 LED connection for local control of an XBee module by a microcontroller.

The boldface characters indicate those typed and sent from the X-CTU Terminal window. This information contains a subtle but important difference between the "D1 4" command sent in an API command packet and the same command sent from the Terminal window. The API packet uses a *hex value* for the parameter you want to change, but the AT command uses the code for the equivalent *ASCII character*. So to change the AD1-DIO1 pin to a logic 0 with an AT command you send the character string "ATD1 4[Enter]" or "ATD14[Enter]". The latter corresponds to the hex values 0x41 0x54 0x44 0x31 0x34 0x0D, where the 0x0D represents the [Enter]. To re-emphasize, the AT command uses the value for the numeral 4 (0x34) while an API command uses the hex value of 4 (0x04).

For this command to work, though, the MCU software must ensure some "quiet time" before it sends the +++. That quiet time, or guard time, comes preset in XBee modules for a 1-second period. After it sends the +++ the MCU must wait for the "OK" and then send the AT command. When an XBee module receives the three plus signs it waits to determine if any other bytes will follow. If none do within the following one second, the XBee module transmits "OK." After that, the MCU can respond with its AT command.

Although the MCU software should wait to receive the "OK" reply before it proceeds, to simplify things the program snippet below for an ARM mbed module simply waits for 1.5 seconds and then transmits the ATD14[Enter] string to the XBee module. In C programs, the "\r" provides a byte equivalent to pressing the Enter key. (You can add similar steps to software for an Arduino Uno module or other MCU.)

```
wait_ms(1500);                  // 1.5 second quiet time

XBeePort.printf("+++");          // send three plus signs

wait_ms(1500);                  // wait 1.5 seconds

XBeePort.printf("ATD14\r");     // transmit ATD0 4 command
```

Now comes another subtle but important point about the AT commands. The XBee modules have an AT Command-Mode Timeout parameter preset for 10 seconds. That setting means after an XBee module receives an AT command, it waits for 10 seconds before it exits the AT command mode. You should ensure no serial communications to the XBee module occur during this time or they might get interpreted as additional AT commands, which you don't want to happen.

During the 10-second AT Command-Mode Timeout period, the state of the AD1-DIO1 pin will not change. The change occurs only after the timeout period ends. So, if you add code to the software for your MCU, do not expect to see an immediate change at the AD1-DIO1 pin. You must wait for the command mode to time out.

You can change the CT – AT Command-Mode Timeout period in the Modem Configuration window for an attached XBee module from the 10-second default value (0x64) in steps of 100 msec. You also can change the GT – Guard Time setting, which has a 1-second default period. Although by convention equipment uses a sequence of three plus signs, +++, to start an AT command period, you may change this character to any other you wish with the CC – Command Sequence Character parameter. I recommend you leave it set for 0x2B, or "+."

It would prove easier to simply turn on an LED at an MCU I/O pin than to control one on an XBee module, but I thought it important to explore AT commands in a bit more depth.

Notes

Note A. In the programs in this experiment, the MCU code expected five characters, so it will wait in the `while (counter < 5)` loop until it receives five characters. Of course, you could set the while loops to receive and transmit more or fewer than five characters, but the software would have the same problem: It would continue to remain "stuck" in the loop until it received all the characters.

You can overcome this problem in two ways:

- Always send the same number of characters. If you don't have enough characters for a transmission, "pad" the end of the message with enough bytes, perhaps 0x00, to give you the proper number to transmit.
- Include a byte count early in a transmission. Remember from earlier experiments that transmissions all started with the same "start" byte, 0x7E, followed by two bytes that indicated the message length. You could use a similar type of format for your own communication protocol.

Note B. The programs in this experiment have a small problem that does not affect operation but could cause confusion if you modify them. The programs all use the same while statement: `while(counter <5)` in two places, first to keep track of the number of received bytes and second to track the number of transmitted bytes. Suppose you decide to receive and transmit six bytes and you change the value in the first while loop: `while(counter <6)`, but you forget to change the next `while` statement, so it stays in the program as: `while(counter <5)`. The program received six characters but transmits only five.

To overcome this problem, use a variable such as `maxbytes` and assign it the value 6 at the start of the program:

```
int maxbytes 6;   //number of bytes to receive and transmit
```

and then set the two `while` statements to:

```
while(counter < maxbytes)
```

Now, one variable, explained with a comment, governs both the transmit and receive counts. And, if you need to know the number of bytes elsewhere in this program, you just use the variable `maxbytes`. You don't have to track down every use of a value "hard coded" in a program.

How to Discover Nearby XBee Modules

Objective: Understand how to discover and identify wireless devices available for communications.

REQUIREMENTS

2 or 3 XBee modules
1 or 2 XBee adapter boards
1 Solderless breadboard
1 USB-to-XBee adapter
1 USB cable—type-A to mini-B
1 Arduino Uno or ARM mbed microcontroller module
1 5-V-to-3.3-V logic-conversion circuit or module (Arduino Uno only)
1 3.3-V-to-5-V logic-conversion circuit or module (Arduino Uno only)
1 1000-ohm (1 kohm), 1/4-watt resistor (brown-black-red)
1 220-ohm, 1/4-watt resistor (red-red-brown)
1 LED
1 Pushbutton (optional)
1 Double-pole double throw (DPDT) toggle switch (Arduino Uno only)
Terminal-emulation software such as HyperTerminal for Windows (ARM mbed only)
Digi X-CTU software running on a Windows PC, with an open USB port

INTRODUCTION

In previous experiments you learned how to use the AT command set and API command packets to communicate with remote XBee modules when you know their addresses. But suppose you do not know the address of a nearby wireless module. Or, suppose someone adds another XBee device to your small network. How can you determine which modules the network can communicate with? You can use an AT command or an API packet to request that all other XBee modules within range respond with their address and name information.

157

The Hands-on XBee Lab Manual.

Table 18.1 Module Information

	Module 1	Module 2	Module 3
MY – 16-Bit Address			
NI – Node Identifier			
SH – Serial Number High			
SL – Serial Number Low			
Note: Module 3 optional.			

In this experiment, you will use two or three XBee modules; one that connects to a microcontroller (MCU) and one or two remote XBee modules that will operate via wireless communications. I used three XBee modules in the following steps. There is no X-CTU configuration file needed for this experiment.

Step 1. To start, I recommend you first use the X-CTU software to restore the factory-default settings. You can do this for each XBee module before you make the configuration changes in the following instructions. Place one of your XBee modules in the USB-to-XBee adapter and click the Restore button in the X-CTU software. Next click on Read to read the module's Modem Configuration information.

- Under the heading Networking & Security, go to the MY – 16-Bit Source Address. If no address exists, or if it shows a zero, click on the label and type in as many as four hexadecimal values. Write your module's MY information in Table 18.1.
- Move down to the last item last item under the Networking & Security heading: NI – Node Identifier and click on this label. Then click on the Set button that appears to the right of this label and type in a name for the module. You can use as many as 20 characters, but I recommend you use four or five characters and meaningful names, such as END, RCVR, MOD1, and so on. Write the NI information for the module in Table 18.1.
- Look under Serial Interfacing heading for the label, AP – API Enable, and click on it. Choose 1 – API ENABLED.
- Write the SH – Serial Number High and, and SL – Serial Number Low information for the module in Table 18.1.
- Finally, click on Write to save this configuration in the attached module. This step enables the API interface on your modules and gives it a name you can recognize.

Repeat Step 1 for each XBee module. All modules in this experiment will have the same configuration.

In this experiment I used three XBee modules, labeled RCVR, XMTR, and END for the NI settings.

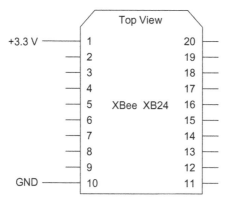

FIGURE 18.1 Power and ground connections for any XBee modules on a solderless breadboard for this experiment. Remove any other connections.

Step 2. Place one XBee module in an adapter on your solderless breadboard. Place your second XBee module in the USB-to-XBee adapter connected to your PC. If you have a third XBee module, place it in a second XBee adapter on the solderless breadboard. For now, the breadboard XBee module or modules require only power and ground, as shown in Figure 18.1. The breadboard modules should not have any other connections.

It doesn't matter which modules you place in XBee adapters on your breadboard, but you must have one module in the USB-to-XBee adapter attached to your PC.

Step 3. Turn on power to the breadboard and ensure you have power at the XBee module on the USB-to-XBee adapter. If not already running, start the X-CTU program. Go to the X-CTU PC Settings window and click on Test/Query to ensure you have an active communication link between the PC and the module in the USB-to-XBee adapter. (Ensure you have set the X-CTU program for the proper COM port used by the USB-to-XBee adapter.)

Step 4. Go to the X-CTU Terminal window and clear the screen. Click on Show Hex to split the Terminal window vertically into a hex-value side (right) and an ASCII-character side (left).

Step 5. The AT command set includes ND—Node Discover, which causes the XBee module attached to the PC to transmit a message that all modules within range will respond to. The XBee modules default to use radio channel C, so all modules will transmit and receive on this channel. If you had configured a module for a different channel, it could not communicate with those set for channel C.

Step 6. Now that you have one or two *remote* XBee modules set up and powered you can issue an ND command to the XBee module attached to your PC to determine what modules it can communicate with.

Go to the X-CTU Terminal window and clear it. Move your cursor into the left column. When told to, you will type +++ to put the USB XBee module into the AT-command mode. The XBee module attached to your PC will reply with OK. After you see the OK, you will type ATND[Enter].

Go ahead and type +++, wait for OK to appear, and then type ATND[Enter]. Do not type anything else and wait a few seconds. What did you observe?

In my X-CTU Terminal window, I saw the information shown in Figure 18.2.

+++OK	2B 2B 2B 4F 4B 0D
ATND	41 54 4E 44 0D
89FA	38 39 46 41 0D
13A200	31 33 41 32 30 30 0D
4049E1E6	34 30 34 38 45 31 45 36 0D
24	32 34 0D
END	45 4E 44 0D
	0D
1AB	31 41 42 0D
12A200	31 33 41 32 30 30 0D
4049E0EC	43 30 34 39 45 30 45 43 0D
2C	32 43 0D
XMTR	58 4D 54 52 0D
	0D
	0D

```
+++OK        2B 2B 2B 4F 4B 0D
ATND         41 54 4E 44 0D
89FA         38 39 46 41 0D
13A200       31 33 41 32 30 30 0D
4049E1E6     34 30 34 39 45 31 45 36 0D
24           32 34 0D
END          45 4E 44 0D
             0D
1AB          31 41 42 0D
13A200       31 33 41 32 30 30 0D
4049E0EC     34 30 34 39 45 30 45 43 0D
2C           32 43 0D
XMTR         58 4D 54 52 0D
             0D
             0D
```

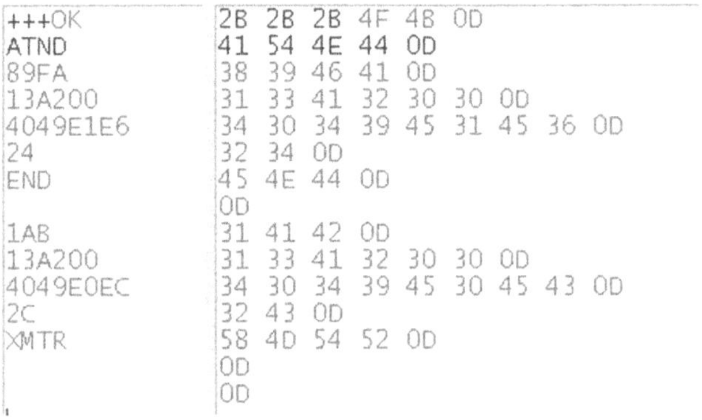

FIGURE 18.2 Data from responses to an ATND command as taken from the X-CTU Terminal window.

If you did not see a response from the XBee module connected to your PC, use the X-CTU Modem Configuration window to read the settings from your attached module. Confirm you have the API set to: (1) API Enable for the XBee module connected to your PC via a USB cable. You do not need the API enabled in modules to discover them, but I recommend you enable API in all modules used in this experiment so they all have the same settings and you can interchange them, if you choose to.

The top two lines in the X-CTU Terminal window show the +++ command sent, the OK received, the ATND command sent, and the hex value 0x0D. The five lines that start with 89FA in the left column indicate:

89FA	= MY – 16-Bit Source Address
13A200	= SH – Serial Number High
4049E1E6	= SL – Serial Number Low
24	= DB – Received Signal Strength
END	= NI – Node Identifier

The next section provides the same types of information from my second remote XBee module, named XMTR:

1AB	= MY – 16-Bit Source Address
13A200	= SH – Serial Number High
4049E0EC	= SL – Serial Number Low
2C	= DB – Received Signal Strength
XMTR	= NI – Node Identifier

Note the information in each line shown in Figure 18.2 ends with 0x0D, the ASCII value for carriage return, which forces the Terminal cursor to move to the line below. Each section for a given XBee module ends with two 0x0D values. Software could detect the single 0x0D and the 0x0D pair and then properly parse, or separate, the MY, SH, SL, DB, and NI information so you could identify individual replying XBee modules. That approach requires a lot of work for two reasons:

- You need the *hex codes*, such as 0x01 and 0xAB, to address a remote module, not the *ASCII bytes* for numerals, such as 0 (0x30), 1 (0x31), A (0x41), and B (0x42), which the ATND command returns.
- The SH value, for example, uses four bytes of information (0013A200), but the response to the ATND command leaves off the leading zeros (13A200), which commands require to address XBee modules.

Because you set all XBee modules for AIP mode, you can get around this problem by sending an API command packet, rather than an ATND command, to the XBee module connected to your computer. The response to the API command returns the needed hex values, and not ASCII character values.

Keep in mind you send the API ND command *to the XBee module connected to your PC*. That module performs the node-discovery operations and replies with information about discovered XBee devices. You do NOT send the ND command to the remote modules.

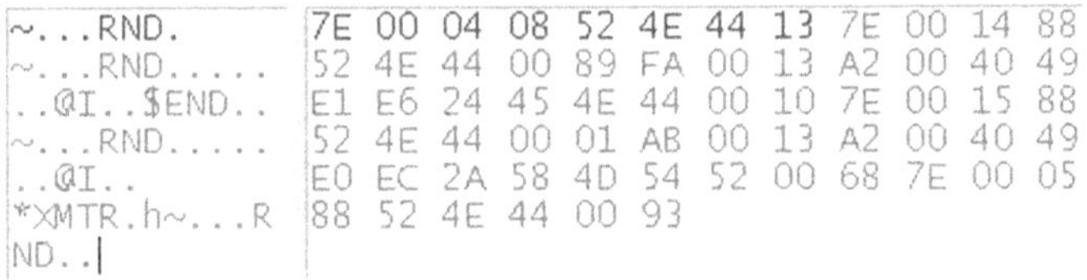

```
~...RND.          7E 00 04 08 52 4E 44 13 7E 00 14 88
~...RND.....      52 4E 44 00 89 FA 00 13 A2 00 40 49
..@I..$END..      E1 E6 24 45 4E 44 00 10 7E 00 15 88
~...RND.....      52 4E 44 00 01 AB 00 13 A2 00 40 49
..@I..            E0 EC 2A 58 4D 54 52 00 68 7E 00 05
*XMTR.h~...R      88 52 4E 44 00 93
ND..|
```

FIGURE 18.3 Responses to an API ND command packet as seen in the X-CTU Terminal window.

Step 7. Clear the X-CTU Terminal window and click on Assemble Packet. Click on the HEX radio button in the lower-right corner of the Send Packet window. Within the Send Packet window, type:

```
7E 00 04 08 52 4E 44 13
```

You have seen this format in previous experiments. The hex information represents:

7E	= start byte
0004	= 4 bytes in message
08	= AT Command Request byte
52	= frame identifier value (all experiments use this value)
4E	= hex code for letter "N"
44	= hex code for letter "D"
13	= checksum for previous four bytes

You do not see MY, SH, or SL information in the packet above because it commands the XBee module directly attached to your PC to perform the ND action. Correct any typing errors. Then, click on Send Data. What do you observe in the Terminal window?

Figure 18.3 shows the information I found for two remote XBee modules, named XMTR and END. You will see different values from your module or modules, but the information will follow the same format. The letter R in the RND information appears because it's the ASCII character equivalent of the 0x52 frame-identifier value used in all experiments.

```
~..RND.           7E 00 04 08 52 4E 44 13 7E 00 14 88

~...RND.....      52 4E 44 00 89 FA 00 13 A2 00 40 49

..@I..$END..      E1 E6 24 45 4E 44 00 10 7E 00 15 88

~..RND.....       52 4E 44 00 01 AB 00 13 A2 00 40 49

..@I..            E0 EC 2A 58 4D 54 52 00 68 7E 00 05

*XMTR.h~...R      88 52 4E 44 00 93

ND..
```

In the top line, you can see the API command packet sent: 7E 00 04 08 52 4E 44 13. Next you find three responses, which always start with 7E:

Response 1:

```
7E 00 14 88 52 4E 44 00 89 FA 00 13 A2 00 40 49 E1 E6 24 45 4E
44 00 10
```

Response 2:

```
7E 00 15 88 52 4E 44 00 01 AB 00 12 A2 00 40 49 E0 EC 2A 58 4D
54 52 00 68
```

Response 3:

```
7E 00 05 88 52 4E 44 00 93
```

Although I used only two remote modules in this experiment, the X-CTU Terminal window displayed three responses, which I'll explain shortly.

You can break down the first response as:

7E	= start byte
0014	= message length (24 bytes)
88	= packet type (Remote AT Command Response)
52	= frame identifier value (all experiments use this value)
4E44	= command name (ND)
00	= status byte (OK)
89FA	= MY – 16-Bit Source Address
0013A200	= SH – Serial Number High
4049E1E6	= SL – Serial Number Low
24	= signal strength
45 4E 44	= NI – Node Identifier (END)
00	= null value 0x00, identifies end of NI characters
10	= checksum for this message

Note the MY, SH, and SL information now provides the proper number of bytes:

MY — 2 bytes, 0x89 and 0xFA

SH — 4 bytes, 0x00, 0x13, 0xA2, and 0x00

SL — 4 bytes, 0x40, 0x49, 0xE1, and 0xE6

Response 3 provides a reply from the XBee module attached to a PC via a USB-to-XBee adapter. It repeats the ND command and includes a status byte that would indicate any error that arises from acting upon the ND command:

0x00 = OK, 0x01 = Error, 0x02 = Invalid Command, and 0x03 = Invalid Parameter.

The final response *does not* indicate the number of XBee modules that responded or that no modules responded. It simply tells you the module attached to your PC properly executed the ND command.

Step 8. As an exercise, use the information in Response 2 (or from your second remote XBee module, if you have one) and parse the data into the proper categories shown here:

     _____   = start byte
     _____   = message length (number of bytes)
     _____   = packet type (Remote AT Command Response)
     _____   = frame identifier value (all experiments use this value)
     _____   = command name (ND)
     _____   = status byte
     _____   = MY – 16-Bit Source Address
     _____   = SH – Serial Number High
     _____   = SL – Serial Number Low
     _____   = signal strength
     _____   = NI – Node Identifier
     _____   = null value 0x00, identifies end of NI characters
     _____   = checksum for this message

Did you get the information you expected from the second remote XBee module? Check the information from your module or modules against the information you wrote earlier in Table 18.1.

Because the Node Identifier (NI) can use as many as 20 bytes, the API response uses a null (0x00) to indicate the end of this string of characters. The byte count *includes* the null value. It does not include the checksum byte, though.

Step 9. When you use an API command, responses from XBee modules provide a consistent number of bytes for the MY, SH, and SL data. Thus software could parse the response from an API node-discovery command to yield useful MY, SH, SL, and NI information for each discovered XBee module.

In the next section you will see how an MCU can issue an API ND command to discover remote XBee modules. For the most part, the code in this experiment does not include error-checking steps that could alert you to a problem transmission or bad data in a response. The program does, though, use the checksum to test for receipt of a bad packet.

When a UART operates at 9600 bits/second, or 104 microseconds/bit, an MCU has about one millisecond to decide what to do with each received byte. So software could take two routes. First, determine what to do with each byte as it arrives, or second, save all the received bytes in an array and process them later.

This experiment assumes you know how many remote XBee modules will respond and it will store the information from each module in an array. In other experiments you will learn how to determine the number of XBee modules in a network and how to use the information "discovered" for each one.

In my lab I had three XBee modules; two connected to power and ground on a solderless breadboard and one connected to power and ground via the USB-to-XBee adapter and a USB cable. In the steps that follow I connected one of the XBee modules on the breadboard to an MCU. The XBee module in the USB-to-XBee adapter remained connected to my PC, but it simply obtained power via the USB connection.

As you examine the code that follows, keep in mind the message-length count in a packet comprises two bytes. So far, API messages and responses have taken fewer than 30 or so bytes. But cases might occur when a message exceeds 0xFF (255) bytes. So a program that receives and processes messages should include the most-significant byte in the byte count. This software multiplies the most-significant byte-count value by 256 and adds it to the least-significant byte-count value to create an accurate byte-count value from 0 to 65,535, although it's unlikely a packet will contain that many bytes.

EXPERIMENT 18 SOFTWARE FLOW CHART

The flow chart in Figure 18.4 shows the operations the code will perform. Descriptions of software operations and instructions for the Arduino Uno and ARM mbed MCU modules follow.

USING AN ARDUINO UNO MODULE

Uno Step 1. The diagram in Figure 18.5 shows the connection from the Arduino Uno module to an XBee module in a breadboard adapter. You can add an optional double-pole double-throw toggle switch to disconnect the logic-level-conversion circuits from the Arduino Uno module, as explained later. If you don't use a switch, wires will suffice. Make these connections now.

Pay careful attention to the notes below and in Figure 18.5 that explain *additional* connections not shown in the diagram for the sake of clarity:

- On LLC 5-to-3, connect a 1-kohm resistor (brown-black-red) between pins 1 and 2.
- On LLC 3-to-5, connect pin 2 to ground.
- On both logic-level-converter devices, connect pins 11, 12, 13, and 22 to ground.
- On both logic-level-converter devices, connect pin 1 to +5 volts.
- On both logic-level-converter devices, connect pins 23 and 24 to +3.3 volts.

Uno Step 2. The circuit for the Arduino Uno module also requires an LED, a pushbutton switch, and two resistors. The LED indicates completion of module discovery and the pushbutton signals the MCU to start the main program steps.

The circuit diagram in Figure 18.6 shows the added LED and pushbutton switch. Make these connections now. You can place the LED and resistors on a solderless breadboard.

Because the Arduino Uno module will use one UART to communicate with an XBee module *and* the host PC USB connection, maintaining the serial connections between an Arduino Uno module and an XBee module when downloading code can cause an electrical conflict. So you MUST manually disconnect (or use a small switch to open) the two serial-port connections between an Arduino Uno module and any external device while the compiler

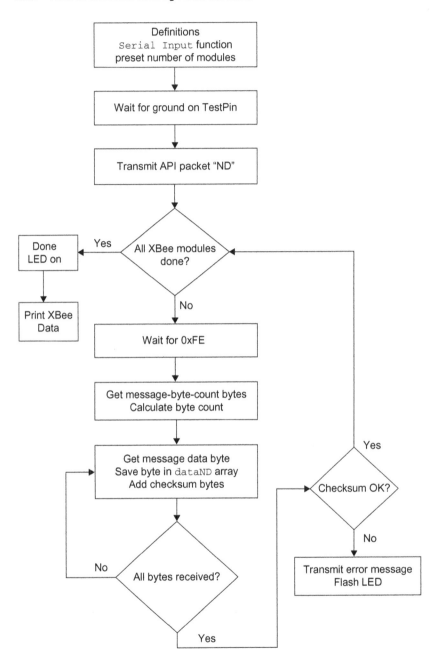

FIGURE 18.4 Flowchart for a program to save XBee module information.

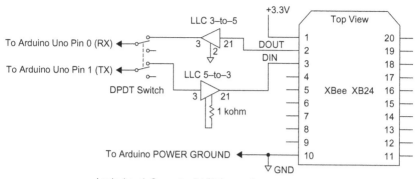

FIGURE 18.5 Connections between an XBee module and an Arduino Uno module require logic-level-converter circuits.

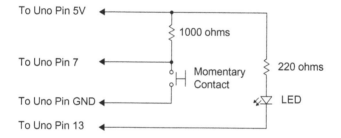

FIGURE 18.6 The Uno module requires these extra components and connections for this experiment.

downloads code to the Arduino Uno board. I used a small toggle switch to disconnect the connections between the logic-level-conversion circuits and the Arduino Uno module during code downloads. Individual wires will work, too.

Arduino Uno Software

The program shown in Program EX18_Uno_ND provides the code used in this section. You can download code examples at: http://www.elsevierdirect.com/companion.jsp?ISBN=9780123914040.

Program EX18_Uno_ND

```
/*
 * EX18 Arduino Uno Module   Rev. D
 * Program EX18_Uno_ND
 * Send ND API command packet to "hub" module and identify set
 * number of modules. Save information in an array and print
 * results as hex data.
 * Jon Titus 08-29-2011
 */

// define API "ND" packet
byte  packetND[] = {0x7E, 0x00, 0x04, 0x08, 0x52, 0x4E, 0x44,0x13};
byte  packetNDlength = 8;

// define number of modules here as a constant, change to suit
const byte numb_of_modules = 2;

// and define the number of bytes in array dataND[]
const byte dataNDlength = 40;
byte  XBee_numb;
byte  dataND[numb_of_modules][dataNDlength];
byte  testdata;

// Define digital I/O pins
int     TestPin = 7;
int     LedPin = 13;

//define other variables
unsigned int  bytecount_hi;
unsigned int  bytecount_lo;
unsigned int  bytecount;
unsigned int  counter;
unsigned int  chksum;
unsigned int chksum_temp;

// function SerialInput reads UART after it has data
byte SerialInput()
{
  while (Serial.available() == 0)       // wait for UART to have
                                        // new data
    {
    }
    return (Serial.read());             // get new data, return
}

// Set up conditions for I/O pins and UART bit rate
void setup()
{
  Serial.begin(9600);
  pinMode(TestPin, INPUT);
  pinMode(LedPin, OUTPUT);
  digitalWrite(TestPin, HIGH);
  digitalWrite(LedPin, HIGH);                  // turn test LED off
                                               // to start
}
```

```
// Main loop for Arduino Uno code. Arduino Uno does not use
// void main()...

void loop()
{
  while (digitalRead(TestPin) == HIGH)     // wait for ground
                                           // contact on TestPin
  {
  }
  delay(500);                              // half-second delay

  Serial.write (packetND, packetNDlength); // transmit "ND" API
                                           // packet to XBee

// This portion of the program gets serial data sent by XBee
// modules and saves the data in an array of bytes for each module.
    for (XBee_numb = 0; XBee_numb < numb_of_modules; XBee_numb++)
    {
      testdata = SerialInput();
      if (testdata == 0x7E);                 // serial
                                             // byte == 0x7E?
      {
        bytecount_hi = SerialInput();
        bytecount_lo = SerialInput();
        bytecount = (bytecount_hi * 256) + bytecount_lo;  // get #
                                             // of bytes in msg
        counter = 0;
        chksum_temp = 0;                               // clear
                                             // checksum value
        for (counter; counter < bytecount; counter++)    // put
                                             // XBee data in
                                             // until done
        {
          dataND[XBee_numb][counter] = SerialInput();
          chksum_temp = chksum_temp + dataND[XBee_numb][counter];
        }
        chksum = SerialInput();                        // get
                                             // checksum--last byte
        if (chksum != (0xFF - (chksum_temp & 0xFF)))   // do
                                             // checksums match
        {
          while(1)                                     // no
                                             // match, flash LED
                                             // forever
          {                                            // error
                                             // handling could go
            digitalWrite(LedPin, LOW);                 // here
                                             // instead.
            delay(1000);
            digitalWrite(LedPin, HIGH);
            delay (500);
          }
        }
```

```
  }                                            // do it
                                               // for next module

  }                                                    // OK,
                                               // got data for all
                                               // XBee modules
digitalWrite(LedPin, LOW);
delay(500);                                    // half-second delay
Serial.print("\n\n\r");                        // go to a new line
                                               // in text window
// Test routine to print data in each XBee array as hex
// characters
// Data goes to PC terminal emulator
  XBee_numb = 0;
  for (XBee_numb; XBee_numb < numb_of_modules; XBee_numb++)
  {
    counter = 0;
    for (counter; counter < 40; counter++)
      {
         if (dataND[XBee_numb][counter] < 0x10)
         {
           Serial.print("0");
         }
       Serial.print(dataND[XBee_numb][counter], HEX);
       Serial.print("  ");
      }

    Serial.print("\n\n\r");                    // go to a new line
                                               // in text window
  }

    while(1)                                   // end program in an
                                               // infinite
                                               // do-nothing
      {                                        // loop
      }
}
// Arduino Uno void loop() code ends here
```

The EX18_Uno_ND program first defines variables. The array packetND[ ] sets up eight bytes for the API ND command. Other definitions set the length of this packet (8), the number of expected modules (2), and the number of bytes (40) allocated for each module's response:

```
byte packetNDlength = 8;

const byte numb_of_modules = 2;

const byte dataNDlength = 40;
```

If you plan to use only one remote XBee module (total of two overall, one remote and one attached to the Arduino Uno), change the value in const byte numb_of_modules = 2; to const byte numb_of_modules = 1; in this experiment, the number of modules refers to those NOT connected to the Arduino

Uno module. Remember, you may leave an XBee module in the USB-to-XBee adapter. It will obtain power from your PC and will not communicate with it.

Instead of repeating serial-input commands throughout the program, the code includes a `SerialInput` function that tests for a byte of unread information at the Arduino Uno UART. As soon as a new byte arrives from an XBee module, the function returns the newest byte to the main loop.

The `setup` section of code establishes the UART data rate at 9600 bits/second, sets the mode for the two MCU I/O pins defined earlier (see Figure 18.6), and sets the initial conditions for those pins. Unlike many MCU compilers that use a `void main()` definition to create the core of a program, the Arduino compiler uses a `void loop()`. The Arduino Uno runs some underlying code, so the loop simply operates "on top" of that code in its own loop.

Within the main loop, the software first waits for a logic 0 at the TestPin input (pin 7). As soon as the code detects this condition it proceeds to the command `delay(500)` that creates a half-second quiet period. Next, the MCU code transmits the eight values in the `packetND` array that holds the API ND-command packet:

```
Serial.write (packetND, packetNDlength);
```

The loop `for   (XBee_numb...` provides the XBee-response processing steps. It waits for the value 0x7E to arrive from the XBee module via the UART. When that value arrives, the code gets the next two bytes—the most-significant byte and then the least-significant byte—that represent the total byte count for the message. The `bytecount` math operation calculates the total number of bytes in the message. Remember, the `bytecount` value does not count the checksum byte.

The next `for` loop takes the arriving bytes and saves them in the array `dataND`, which can store as many as 40 values, and then adds the received bytes to calculate a checksum. When the routine receives the checksum byte from the XBee module, the code compares it with the calculated checksum. If the checksums do not match, the code branches into a short loop that continuously pulses the LED shown previously in Figure 18.6.

After the program receives all the values from the preset number of remote XBee modules, it displays the hex values of the remote-module information so you can easily review them. The version of the C language for the Arduino Uno module does not include a command that formats hex values with leading zeros. So an `if` statement checks for hex values from 0x00 to 0x0F and gives then a 0 prefix in the Arduino Serial Monitor window. Thus you will see results such as `AB 00 13 A2 00` rather than `AB 0 13 A2 0`.

Uno Step 3. If you have not included a DPDT switch in your Arduino Uno circuit as shown in Figure 18.5, you must label the two wires that connect to the two Uno serial-port pins, TX and RX. Temporarily disconnect these wires from the Uno module's serial-port pins. (Or flip the switch to disconnect these signals.) You can simply remove the wires from the female connector strip and reinsert them later.

Load the program EX18_Uno_ND into the Arduino Uno compiler, compile it, and correct any errors. When the compiler completes loading the code into the Arduino Uno module, as indicated by the Done Saving message in the bottom text area, reconnect the wires (or flip the switch) from the logic-level-converter circuit to the serial-port pins on your Uno module. Now you have the Arduino Uno module reconnected to the logic-level-conversion circuits and the XBee module.

Arduino Uno Troubleshooting

When you compile and load a program into an Arduino Uno MCU, do not leave the logic-level-converter circuits, or any other device, connected to the Arduino Uno serial-port pins, RX <--0 and TX→1. Signal conflicts at the MCU UART input can cause the compiler to display the error message "Problem uploading to board." Don't worry; just temporarily disconnect the two signal wires from the logic-level-conversion circuits to the Arduino Uno module and recompile and download the program code. When the compiler indicates: Done uploading, reconnect the wires or flip the switch.

Uno Step 4. After you successfully compile the EX18_Uno_ND program and it has loaded into your Arduino Uno, press the Uno reset button. An LED on the Arduino Uno will flash several times and turn off. Click on the Arduino Serial Monitor button to open its window.

Either press the pushbutton shown in Figure 18.6 or make a quick ground connection to the Uno-Pin-7 end of the 1000-ohm resistor. Grounding this connection lets the EX18_Uno_ND program proceed.

As soon as the MCU has processed the information from the remote XBee module or modules, the LED shown in Figure 18.6 will turn on. (Don't confuse this LED with the small LEDs on the Uno module.) At this point, you should see characters, such as:

```
~...RND.
```

in the Serial Monitor window. These characters represent the eight bytes in the transmitted API ND-command packet. In a second or two you also should see information from your XBee modules—40 bytes per module—that looks like:

```
88 52 4E 44 00 12 34 00 12 A2 00 40 49 E0 28 2D 52 43 56 52 00

00 00 ...


88 52 4E 44 00 89 FA 00 12 A2 00 40 49 E1 E6 1D 45 4E 44 00 00

00 00 ...
```

The C code displays all 40 values in each `dataND` array for an XBee module because the software defined each array with 40 elements. The information matched what I expected based on the data shown earlier in Figure 18.3 for my two remote modules.

You might wonder why you see the ~...RND., or similar information, at the start. The Arduino Uno module has only one UART output, so the Arduino Serial Monitor "sees" the 8-byte ND command sent to the XBee module attached to the Arduino Uno as well as the XBee information the MCU transmits to the Serial Monitor. Not all hex codes in the API packet represent printing ASCII characters, thus the odd symbols. This "mixing" of API-command and XBee-response data illustrates the problem of sharing a UART between two external devices.

Remember, the EX18_Uno_ND software identified the one or two XBee modules with only power and ground connected to them. The "//Test rou-tine to print data..." portion of the EX18_Uno_ND program formats the bytes so they appear as hexadecimal characters. Then the program ends in an infinite while(1) loop.

Arduino Uno Conclusion

In this experiment you learned how to use an API packet with the ND command to request identification information from modules on the same default radio channel. You also learned the responses from XBee modules provide useful information. In this experiment, an Arduino Uno MCU served as the "hub" for communications with other XBee modules.

Because the program saved the information from each XBee module in an ordered fashion, as shown earlier in Step 7 in the first part of this experiment that did not involve an MCU, you can locate information at specific places in the dataND array. Thus the XBee SH address for XBee module 0 in the dataND array exists in array elements dataND[0][7] through dataND[0][10], and the SL value exists in array elements dataND[0][11] through dataND[0][14].

To simplify retrieving information for a given XBee module, and give others insight into what you have programmed, you might use a structure in a C program to store the replies from XBee modules. This structure declaration sets up arrays of bytes for XBee data MY (2 bytes), SH (4 bytes), SL (4 bytes), and NI (20 bytes plus a null). You could expand it to hold other information, too.

```
struct XBeedata
{    byte MY[2];
     byte SH[4];
     byte SL[4];
     byte DS[1]
     byte NI[21]
} XBee[2];
```

The XBee[2] at the end of the structure definition creates two structures, XBee[0] and XBee[1], for two remote modules. Then the loop that obtains replies from XBee modules could put the information in the structure XBee[0] or XBee[1] as the bytes arrive from the UART.

Please skip ahead to the Final Steps section.

USING AN ARM MBED MODULE

mbed Step 1. The diagram in Figure 18.7 shows the connections from the ARM mbed module to an XBee module. Make these connections now. The ARM mbed module does not require logic-level-conversion circuits because it operates at 3.3 volts, which makes it compatible with the logic levels on an XBee module.

mbed Step 2. The circuit for the ARM mbed module requires a few added components: an LED, a pushbutton switch, and two resistors. The LED indicates completion of module discovery and the pushbutton signals the MCU to start the main program steps.

The circuit diagram in Figure 18.8 shows the added LED and pushbutton switch. Make these connections now. You can place the LED and resistors on a solderless breadboard.

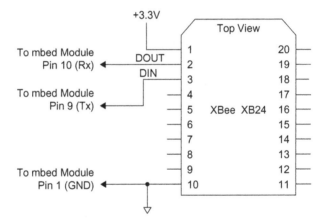

FIGURE 18.7 An ARM mbed module operates with 3.3-volt logic, so it can connect directly to an XBee module.

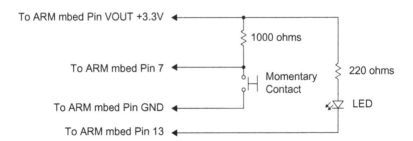

FIGURE 18.8 The ARM mbed module requires these extra components and connections for this experiment.

ARM mbed Software

The program shown in EX18_mbed_ND provides the code used in this section. You can download code examples at: http://www.elsevierdirect.com/companion.jsp?ISBN=9780123914040.

The EX18_mbed_ND program first defines variables. The array `packetND[ ]` sets up eight bytes for the API ND command, sets the length of this packet (8), the number of expected modules (2), and the number of bytes (40) allocated for each module's response:

```
int packetNDlength = 8;

const int numb_of_modules = 2;

const int dataNDlength = 40;
```

If you plan to use only one remote XBee module (total of two overall, one remote and one attached to the ARM mbed), change the value in `const int numb_of_modules = 2;` to `const int numb_of_modules = 1;` in this experiment, the number of modules refers to those NOT connected to the ARM mbed module. Remember, you may leave an XBee module in the USB-to-XBee adapter. It will obtain power from your PC and will not communicate with it.

The ARM mbed serial ports default to 9600 bits/second, so there's no need to specify this bit rate in the code. Serial port `Sport` connects to pin 9 (MCU UART transmitter) and pin 10 (MCU UART receiver). A second, predefined MCU serial port, `pc`, connects to your host PC as a virtual serial port via the module's USB connection. The two UARTs operate independently of one another. Thus, the `pc` serial connection lets you use a terminal-emulator program to display information separate from the MCU serial port at pins 9 and 10 that connects to an XBee module.

Instead of repeating serial-input commands throughout the program, the code uses a `SerialInput` function to test for a new byte of data at the serial input and then return the byte. As soon as a new byte arrives from an XBee module, the function returns the newest byte to the main loop.

Within the main loop, the software first waits for a logic 0 at the TestPin input (pin 7). As soon as the code detects this condition it proceeds to the command `wait(0.5)` that creates a half-second quiet period. Then it proceeds to the statements:

```
counter = 0;

while (counter < packetNDlength)

    {

    Sport.putc(packetND[counter]);

    counter = counter +1;

    }
```

This loop transmits the eight values in the `packetND` array to the `Sport` UART.

The loop `for (XBee_numb...` provides the XBee-response processing steps. It waits for the value 0x7E to arrive from the serial port. When an 0x7E value arrives, the code gets the next two bytes—the most-significant byte and then the least-significant byte—that represent the total byte count for the message. The `bytecount` math operation calculates the total number of bytes in the message, 0 to 65,535. Remember, the `bytecount` value does not count the checksum byte. (A response from an XBee module in API mode always starts with the hex value 0x7E.)

The next `for` loop takes the arriving bytes and saves them in the array `dataND`, which can store as many as 40 elements, and then adds the received bytes to calculate a checksum. When the routine receives the checksum byte from the XBee module it compares it with the calculated checksum. If the checksums do not match, the code branches into a short loop that continuously pulses the LED shown previously in Figure 18.8.

After the program receives all the values from the preset number of remote XBee modules, it displays the hex values of the information so you can easily review them.

ARM mbed Program EX18_mbed_ND.cpp

```
/*
 * EX18 ARM mbed Module   Rev. E
 * Program EX18_mbed_ND
 * Send ND API command packet to "hub" module and identify set
 * number of modules. Save information in an array and print
 * results as hex data.
 * Jon Titus 08-29-2011
 */

#include "mbed.h"

// Define API "ND" command-packet array here
int  packetND[] = {0x7E, 0x00, 0x04, 0x08, 0x52, 0x4E, 0x44, 0x13};
int  packetNDlength = 8;

//Define number of modules here as a constant
const int  numb_of_modules = 2;

//and define the number of bytes in array dataND[] for XBee modules
const int dataNDlength = 40;
int  dataND[numb_of_modules][dataNDlength];

//Input-output and serial I/O definitions
DigitalIn   TestPin(p7);
DigitalOut  LedPin(p13);
Serial Sport(p9, p10);
Serial pc(USBTX, USBRX);
```

```
//Define additional variables
int   XBee_numb;
int   testdata;
int   MSbits, LSbits;
unsigned int  bytecount_hi;
unsigned int  bytecount_lo;
unsigned int  bytecount;
unsigned int  counter;
unsigned int  chksum;
unsigned int  chksum_temp;

//Function SerialInput reads UART after it has new data
int SerialInput()
{
  while (Sport.readable() == 0)        //wait for UART to have
                                       //new data
  {
  }
    return (Sport.getc());             //get new data, return
}


//Main XBee program


int main(void)
{
  LedPin = 1;               //turn test LED off to start
  while (TestPin == 1)      //wait for ground contact on TestPin
  {
  }
  wait(0.5);                //half-second delay

  //Transmit packetND
  counter = 0;
  while ( counter < packetNDlength)
  {
  Sport.putc(packetND[counter]);
  counter = counter + 1;
  }

  //This portion of the program gets serial data sent by XBee
  //modules and saves the data in an array of bytes for each module.

  for (XBee_numb = 0; XBee_numb < numb_of_modules; XBee_numb++)
    {
    testdata = SerialInput();
    if (testdata == 0x7E); //first serial byte == 0x7E?
      {                         //yes
      bytecount_hi = SerialInput();
      bytecount_lo = SerialInput();
      bytecount = (bytecount_hi * 256) + bytecount_lo;
                        //get # of bytes in msg
```

```
chksum_temp = 0;                //clear checksum value
        for (counter = 0; counter < bytecount; counter++)
                                //put XBee data in array
        {
        dataND[XBee_numb][counter] = SerialInput();
        chksum_temp = chksum_temp + dataND[XBee_numb][counter];
        }
         chksum = SerialInput();
                                //get checksum--last byte
        if (chksum != (0xFF - (chksum_temp & 0xFF)))
                                //do checksums match
           {
           while(1)
                                //NO MATCH, flash LED forever
              {
                                //Error handling could go
              LedPin = 0;   //here instead.
              wait(1);
              LedPin = 1;
              wait(0.5);
              }
           }
        }                       //YES MATCH, do it for next module

  }
                //OK got data for all XBee modules
  LedPin = 0;
                //turn on LED
  wait(0.5);      //half-second delay

  //Routine to print data in each XBee array as hex characters
  //Data goes to PC terminal emulator
  for (XBee_numb = 0; XBee_numb < numb_of_modules; XBee_numb++)
  {
  pc.printf("\n\r");
      for (counter = 0; counter < dataNDlength; counter++)
      {
        pc.printf("%02X  ", dataND[XBee_numb][counter]);
      }
      pc.printf("\n\r");    //go to a new line in text window
  }

  while(1)                      //end program in an infinite
                                //do-nothing loop
  {
  }
}
```

mbed Step 3. After you load the program EX18_mbed_ND into the ARM mbed compiler, compile it and correct any errors. When you have an error-free program, you will see Success! appear in the Compiler Output area at the bottom of the compiler window. You will also see a file-download window appear. You want to save the compiler output, so click Save. In the Save As window, choose the ARM mbed "drive." (The ARM mbed module looks like a USB memory stick to your PC. It operated as the F: drive on my lab PC.) Then save the compiler output on the ARM mbed. If necessary, you can overwrite a program with the same name on the ARM mbed module. Someone else might have used the ARM mbed, or you might have revised or corrected a program with the same name several times. An overwrite is OK.

mbed Step 4. Start your terminal emulator program and ensure it connects to the virtual serial port that links the ARM mbed to your PC. On a Windows PC, use the Device Manager to find the proper serial port, labeled ARM mbed Serial Port. Check the emulator settings to ensure they match 9600 bits/second, eight data bits, no parity, one stop bit (8N1, and no flow control).

mbed Step 5. Turn on power to your breadboard and press the ARM mbed reset button. The module's blue LED will flash several times and then turn on. Press the pushbutton shown in Figure 18.8 or make a brief contact between the ARM mbed-pin-7 end of the 1000-ohm resistor and ground. DO NOT try to touch ground to pin 7 on the mbed module itself. You risk damaging the MCU.

As soon as the ARM mbed MCU has processed the information from the remote XBee modules, the external LED will turn on. Next you should see information in the terminal emulator—40 bytes per reply—that looks something like:

```
88 52 4E 44 00 12 34 00 12 A2 00 40 49 E0 28 2D 52 43 56 52 00

00 00 ...



88 52 4E 44 00 89 FA 00 12 A2 00 40 49 E1 E6 1D 45 4E 44 00 00

00 00 ...
```

My code displayed all 40 values in each of the two `dataND` arrays because I defined each array with 40 elements. This information matched what I expected based on the information in Figure 18.3 for my two remote modules.

Remember, you have identified the XBee module with only power and ground and any XBee module attached to the USB-to-XBee adapter that connects to your PC.

The "`//Test routine to print data...`" portion of the program formats the bytes so they appear as hexadecimal characters. Then the program ends in an infinite `while(1)` loop.

ARM mbed Conclusion

In this experiment you learned how to use an API packet with the ND command to request identification information from modules on the same default radio channel. You also learned the responses from XBee modules provide useful information. In this experiment, an ARM mbed MCU served as the "hub" for communications with other XBee modules.

Because the program saved the information from each XBee module in an ordered fashion, as shown earlier in Step 7 in the first part of this experiment that did not involve an MCU, you can locate information in specific elements in the `dataND` array. Thus the XBee SH address for XBee module 0 in the `dataND` array exists in array elements `dataND[0][7]` through `dataND[0][10]`, and the SL value exists in array elements `dataND[0][11]` through `dataND[0][14]`.

To simplify retrieving information for a given XBee module, and give others insight into what you have programmed, you might use a structure in a C program to store the replies from XBee modules. This structure declaration sets up arrays of bytes for XBee data MY (2 bytes), SH (4 bytes), SL (4 bytes), and NI (20 bytes plus a null). You could expand it to hold other information, too.

```
struct XBeedata
{    char MY[2];
     char SH[4];
     char SL[4];
     char DS[1];
     char NI[21]
} XBee[2];
```

The XBee[2] at the end of the structure definition creates two structures, XBee[0] and XBee[1], for two remote modules. Then the loop that obtains replies from XBee modules could put the information in the structure XBee[0] or XBee[1] as the bytes arrive from the UART. By using the structure elements XBee[0].MY[0] and XBee[0].MY[1] you know right away a program uses the first and second bytes of the MY address for the XBee 0 module in your network.

Please continue with the next section.

FINAL STEPS

This experiment assumed you knew how many remote modules you plan to communicate with and that you used this number as the numb_of_modules constant in the program. But in many situations you might have from one to, say, 10 modules. How would you create a program to determine when all remote modules had responded to a node-discovery (ND) command?

The XBee modules provide a possible solution: When you assembled and sent the API packet for the ND command, the response included information from the powered remote modules, as well as the "extra" information as noted in Step 7 in the first part of this experiment that did not involve an MCU. This extra information takes the form:

```
7E 00 05 88 52 4E 44 00 93
```

7E	= start byte
0005	= message length (5 bytes)
88	= packet type (Remote AT Command Response)
52	= frame identifier value (all experiments use this value)
4E44	= command name (ND)
00	= status byte (OK)
93	= checksum for this message

From this response you only know the hub XBee module properly executed and completed the ND command operations (status byte = 0x00). You do not know whether the hub received any responses or the number of responses received.

In addition to processing the information about remote modules, your code also could examine incoming data for the packet shown above. Perhaps if you received a message with a byte count of 0x0005, a program could check the status byte, leave the data-gathering routine, and proceed with other parts of your program. I don't recommend that approach, though.

In a later experiment you will learn how to more easily overcome the problem of discovering and counting modules when you have an unknown number of them.

IMPORTANT NOTES

*   The code in this experiment used a variable, `packetNDlength`, to specify the preset length of the ND command packet. You could have used a `sizeof` function to determine the number of elements in the array. But working with arrays in C can get tricky, so I prefer to use a fixed value rather than a calculation. And don't try to calculate an array length when you "pass" an array to a function, because the compiler only gives the function a pointer to the array and not the complete array to work with.

*   The ND command has two related configuration settings, NO – Node Discovery Options and NT – Node Discovery Time.

The Node Discovery Option lets you decide whether or not to have the hub XBee module that discovers other modules also "discover" itself. If you enable this option (NO = 1), you will receive a complete packet of information just as you would for a discovered remote XBee module. I recommend you leave the NO set to its factory-default setting, 0, which disables the option.

The Node Discovery Time lets you set a value from 100 milliseconds (0x01) to 25.2 seconds (0xFC) during which the hub XBee module will wait for remote XBee modules to respond. The NT setting has a default value of 0x19, or 2.5 seconds. I recommend you leave the NT setting at its factory-default value. Multiply the NT value by 100 msec to obtain the delay period.

During this period, remote XBee modules can respond. At the end of this period, when we expect all remote XBee modules have responded, the hub XBee module responds with the information shown earlier in this Final Steps section.

How to Set Up a Small XBee Network

Objective: Understand how to associate modules in a network as they turn on, how to "take attendance" of associated End-device modules, how to address modules by name, and how to detect network errors. This experiment includes critical information about how to reset a Coordinator module.

REQUIREMENTS

2 or 3 XBee modules
1 or 2 XBee adapter boards
1 Solderless breadboard
1 USB-to-XBee adapter
1 USB cable—type-A to mini-B
2 LEDs
2 330-ohm, 1/4-watt, 10% resistors (orange-orange-brown)
Digi X-CTU software running on a Windows PC, with an open USB port

INTRODUCTION

In previous experiments you learned how to communicate with remote XBee modules and query them to determine serial numbers, obtain information about digital and analog signals, and exchange other information. In this experiment you will learn how to associate XBee modules in a *non-beacon network* that comprises one Coordinator module and one or more End-device modules. The diagram in Figure 19.1 shows the arrangement of two End devices and a Coordinator module.

Until now, experiments assumed no other wireless devices in your vicinity would operate on the same wireless channel assigned by default to XBee modules. The factory settings cause each module to operate on the C channel, one of 16 between 2.405 and 2.480 gigahertz (GHz). The C channel (0x0C) operates at 2.410 GHz. But other XBee modules or wireless devices that comply with the IEEE 802.15.4 wireless standard could operate on that channel, too.

183

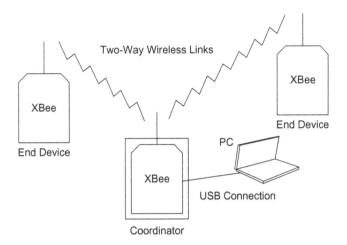

FIGURE 19.1 In a non-beacon network, one XBee module acts as the Coordinator and other XBee devices act as End devices. Communication between End devices must go through the Coordinator.

To avoid communication conflicts, in a non-beacon network of XBee modules, a Coordinator can perform three network-related operations:

- Scan all 16 wireless channels to find one with no radio signals, or one with the smallest amount of radio-frequency (RF) energy. It will use this channel for communications with End-device modules.
- Reassign the Personal-Area Network (PAN) number, which has a factory default of 3332. The Coordinator should not use a PAN value already assigned to a nearby network, so it will test channels to determine PAN values for existing networks within its range and will then choose an unused PAN value.
- Let any XBee device associate with it, or not let XBee devices associate with it. In almost every case, you will allow association unless you have an established network and do not expect to add any extra modules or do not want rogue modules to associate with a network. After you establish a network, you might not want to allow any other XBee devices to enter the network after it reaches a limit of, say, 12 End devices.

End devices also can operate in several ways. They can:

- Associate with any Coordinator with any PAN ID, or only with a Coordinator that has the same preset PAN ID.
- Associate with a Coordinator that operates on any wireless channel, or only with a Coordinator on the same channel.
- Block any association attempts, or try to associate with a Coordinator until it can form an association with one. In other words, continue to try to associate indefinitely.
- Ask the associated Coordinator for any pending information the End device should receive after it comes out of a sleep mode, or do not ask for any information upon awaking.

In this experiment, you will set up one Coordinator module that will:
* Find an unused wireless channel, and
* Let any End device associate with it.

You will set up one or more End-device modules that will:
* Associate with a Coordinator on any channel, and
* Attempt to associate with a Coordinator indefinitely.

The Coordinator will remain attached to your host PC via the USB-to-XBee adapter and USB cable, and you will communicate with the Coordinator via the X-CTU software. The End-device module or modules will operate on their own. I used four XBee modules in the following steps: three modules operated as End devices and one served as the Coordinator. The Coordinator and End modules require different configuration steps.

In the following steps, remember that modules called a Coordinator or an End device remain standard XBee modules. You simply configure them for specific functions, and you can always reconfigure with the factory-supplied settings.

Step 1. End Module or Modules: If you don't know the configuration of your XBee module, I recommend you restore it with the factory-default values. Within the X-CTU window, click on the Modem Configuration tab and in turn place each module in the USB-to-XBee adapter and click on Restore under the Modem Parameters and Firmware heading. You will find the Modem Configuration profile for the End-device modules in the file EX19_End.pro.

Place one of your designated End-device XBee modules in the USB-to-XBee adapter and use the X-CTU software to Read its Modem Configuration information.
* Under the heading, Networking & Security, go to the MY – 16-Bit Source Address. If no address exists, or if it shows a zero, click on the label and type in as many as four hexadecimal values. If a MY value exists, you can use it as is or type in a new Source Address.
* Ensure you have a value of 0 for CE – Coordinator Enable and a value of 0x1FFE for SC – Scan Channels.
* For the A1 – End Device Association, select the setting 6 – 0110B, where the B stands for binary. This setting establishes the conditions: associate with a Coordinator on any channel, and attempt to associate with a Coordinator indefinitely.
* Move down to the last item under the Networking & Security heading: NI – Node Identifier and click on this label. Next, click on the Set button that appears to the right of this label and type in a name for the module. You can use as many as 20 characters, but I recommend you use only four or five.
* Look under the Serial Interfacing heading for the label, AP – API Enable, and click on it. Choose 1 – API ENABLED.
* In Table 19.1, write the name (NI) of each End-device module and its MY, SH, and SL information. You will need this information later so you can compare it with responses from your modules.
* Finally, click on Write to save this configuration in the End-device module. This step enables the API interface on your modules, gives it a name you

Table 19.1 End-Device Module Information

	End-Device 1	End-Device 2	End-Device 3
MY – 16-Bit Address			
NI – Node Identifier			
SH – Serial Number High			
SL – Serial Number Low			
Note: End-device modules 2 and 3 are optional.			

can recognize, and sets it as an End device. (I named my three End-device modules XMTR, PRO, and END.)

Repeat Step 1 for each XBee module you will use as an End device and set aside these modules.

Step 2. Coordinator Module: Place the XBee module designated as your Coordinator in the USB-to-XBee adapter and use the X-CTU software to Read its Modem Configuration information. You will find the Modem Configuration profile for the Coordinator modules in the file EX19_Coord.pro.

- Under the heading, Networking & Security, go to the MY – 16-Bit Source Address. If no address exists, or if it shows a zero, click on the label and type in as many as four hex values. If a MY value exists, you can use it as is or type in a new Source Address.
- Ensure you have a value of 0x1FFE for SC – Scan Channels.
- In the line labeled CE – Coordinator Enable, select 1 – COORDINATOR which lets the Coordinator find an unused wireless channel, and lets any End device associate with it.
- For the A2 – Coordinator Association setting, select 6 – 110B, where the B stands for binary. This setting establishes the conditions: find an unused wireless channel and let any End device associate with this Coordinator.
- Move down to the last item under the Networking & Security heading: NI – Node Identifier and click on this label. Next, click on the Set button that appears to the right of this label and type in a name for the module. You can use as many as 20 characters, but I recommend you use four or five.
- Look under the Serial Interfacing heading for the label, AP – API Enable, and click on it. Choose 1 – API ENABLED.
- Finally, click on Write to save this configuration in the attached module. This step enables the API interface on your modules, gives it a name you can recognize, and sets it as an End device. (I gave my Coordinator the name RCVR.)

Step 3. Leave the Coordinator module in the USB-to-XBee adapter. Place the End-device module or modules in separate XBee adapters on the same or separate breadboards. I recommend placing no more than two XBee modules and adapters in one solderless breadboard.

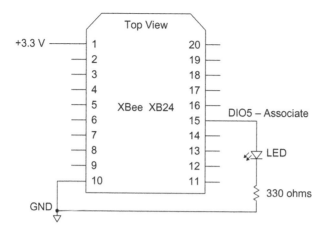

FIGURE 19.2 Power, ground, and LED connections for each XBee module used as an End device.

Each End-device module requires only power, ground, and an LED connection as shown in Figure 19.2. Make these connections now. The LED serves as an Association indicator to let you know an End-device module has properly associated with a Coordinator. You should not have any other connections on End-device modules at this time.

Note: An End-device XBee module does not require an Associate LED. It simply serves as a visual indicator for testing and debugging. Because an LED consumes power, you might eliminate it in battery-powered End devices. If you don't need the Association LED, you can use the DIO5 pin for analog inputs or digital I/O signals.

Step 4. Ensure you have the Coordinator module connected to your host PC via a USB cable. Start the X-CTU program. Go to the X-CTU PC Settings window and click the Test/Query button to ensure you have an active communication link with your Coordinator module.

Step 5. Turn on power to your End-device modules. Each LED should turn on and stay on. After about three seconds, the LEDs should start to flash. One End-device module will start to flash its LED first because it associated with the Coordinator first. The LED on each associated End-device module will continue to flash for as long as you have the module powered.

If an End-device module LED does not turn on when you apply power, recheck your connections and the polarity of the LED. If after five or six seconds the LED still does not flash, recheck the module's Modem Configuration settings. Also check to ensure you are not "off by one pin" when you inserted the module in the breadboard adapter and that you did not plug in a module backwards.

Step 6. Go to the X-CTU Terminal window and clear the screen. If the hexadecimal column is open, you may close it. Type in the $+++$ characters and wait for the OK reply from the Coordinator attached to your host PC.

Type in ATND[Enter], which you used in Experiment 18 to perform a node-discovery operation. You should see information for each associated End-device module. In my Terminal window, I saw the information shown next:

```
+++OK

ATND

FFFE            FFFE            FFFE

13A200          13A200          13A200

4049E0EC        4049E1E6        406AF5AB

2C              26              4B

XMTR            END             PRO
```

Your results should look similar, depending on how many End-device XBee modules you set up. The results for one End device indicate:

FFFE	MY – 16-Bit Source Address
13A200	SH – Serial Number High
4049E0EC	SL – Serial Number Low
2C	DB – Received Signal Strength
XMTR	NI – Node Identifier

The 0xFFFE value for the 16-bit source address (MY) seems unexpected. You probably set the MY parameter to some other 2-byte value, such as 0x89AB or 0x1234, and expected to see it instead of FFFE. According to Digi's manual for the XBee and XBee PRO modules:

> When an End Device associates to a Coordinator, its MY parameter
> is set to 0xFFFE to enable 64-bit addressing. The 64-bit address of
> the module is stored as SH and SL parameters. To send a packet to a
> specific module, the Destination Address (DL+DH) on the sender must
> match the Source Address (SL+SH) of the desired receiver.

The 0xFFFE MY setting holds true for as long as you have an End device associated with a Coordinator. If you later use the X-CTU program to read the Modem Configuration information from an End-device module not associated with a network, you will find the original 16-bit MY value you set earlier in Step 1. Thus, an MY setting of 0xFFFE represents a *temporary* value for associated End-device modules.

Step 7. Clear the X-CTU Terminal window, click on Show Hex, and click on Assemble Packet. In the Send Packet window, go to the space labeled Display, click on HEX. Within the Send Packet window, type:

```
7E 00 04 08 52 4E 44 13
```

This set of hex values represents the API ND command packet used to discover XBee modules. You can use this command to discover modules in a network, too.

Check your typing and correct any errors. Then click on Send Data. What do you observe in the X-CTU Terminal window? Figure 19.3 shows what I observed for my three End-device modules, XMTR, PRO, and END.

```
~...RND.        7E 00 04 08 52 4E 44 13 7E 00 15 88
~...RND.....    52 4E 44 00 FF FE 00 13 A2 00 40 49
..              E0 EC 37 58 4D 54 52 00 0A 7E 00 14
@I..7XMTR..     88 52 4E 44 00 FF FE 00 13 A2 00 40
~...RND.....    49 E1 E6 2B 45 4E 44 00 8F 7E 00 05
..@I..+END..    88 52 4E 44 00 93
~...RND..
```

FIGURE 19.3 Responses to an API ND command packet as seen in the X-CTU Terminal window for two associated End devices. Boldface type helps separate the four response packets.

```
~...RND.        7E 00 04 08 52 4E 44 13 **7E 00 15 88**

~...RND.....    **52 4E 44 00 FF FE 00 13 A2 00 40 49**

..@I..          **E0 EC 40 58 4D 54 52 00 01** 7E 00 14

@XMTR..         88 52 4E 44 00 FF FE 00 13 A2 00 40

~...RND.....    6A F5 AB 4B 50 52 4F 00 5B **7E 00 14**

..@j..KPRO.     **88 52 4E 44 00 FF FE 00 13 A2 00 40**

[~...RND....    **49 E1 E6 32 45 4E 44 00 88** 7E 00 05

...             88 52 4E 44 00 93

@I..2END..

~...RND..
```

Step 8. Use the template that follows to separate bytes in one packet shown in Figure 19.3 into useful information:

_____ = start byte

_____ = message length (number of bytes)

_____ = packet type (Remote AT Command Response)

_____ = frame identifier value (all experiments use this value)

_____ = command name (ND)

_____ = status byte

_____ = MY – 16-Bit Source Address

_____ = SH – Serial Number High

_____ = SL – Serial Number Low

_____ = signal strength

_____ = NI – Node Identifier

_____ = null value 0x00, identifies end of NI characters

_____ = checksum for this message

The fourth message has a slightly different format that provides status information from the Coordinator module.

WHY ASSOCIATE XBEE MODULES IN A NETWORK?

You might wonder, "If I can get the same information from remote modules by using an API ND command packet, why go through the steps to form a non-beacon network of modules?" By creating a network with many End devices and one Coordinator you avoid problems that could disrupt module-to-module communications.

A module configured as a Coordinator can perform an energy scan to find an unused channel. Without this capability you would have to assign each module to the same wireless channel and hope no nearby networks use that channel. This sort of "keep-your-fingers-crossed" approach might work in a lab, but many devices now operate within the same frequency span as the XBee modules, so letting a Coordinator seek an open channel for a network provides a better result. Likewise, a Coordinator can find an unused personal-area network (PAN) ID and use it to keep one network separate from others nearby.

A non-beacon network of XBee modules also lets you "take attendance" quickly to determine if specific End-device modules have associated with the Coordinator. This technique uses a new command, DN – Destination Node.

Step 9. You should have written the NI – Node Identifier for one or more End-device modules in Step 1, Table 19.1. Now you can use those End-device module names to determine if they still exist in your network.

With your network powered and operating—the End-device LEDs should continue to flash—go to the X-CTU Terminal window and clear the screen. Now you will get ready to send the AT command ATDN[your_module_name_here][Enter] to the Coordinator. You could send this command in an API packet, but for now the stand-alone AT command will work.

Refer to Table 19.1 and write the names (NI – Node Identifier information) of your End-device modules in the space below.

Again clear the X-CTU Terminal window and have the window set to display hex values. In the left-hand column, type +++ and wait for the OK reply. Next, type ATDN followed immediately by the name you assigned one of your End-device modules. (I typed ATDNXMTR[Enter].) Be sure to use the same upper- or lower-case characters used as your End-device Node Identifier.

When I performed this step, I saw:

```
+++OK
```

```
ATDNXMTR
```

```
OK
```

So I know the End-device module named XMTR still operates in the network, associated with my Coordinator module. Suppose you type the name of a nonexistent End device after the ATDN. What would happen? Try it. You should see something like:

```
+++OK
```

```
ATDNtest
```

```
ERROR
```

If you have a network with several End devices, you could write an MCU program to transmit DN commands in an API packet with the names of End devices to quickly ensure all expected End-device modules exist and can operate. Your program might use a table of Node Identifier names that it would send to the Coordinator and then wait for a response of either OK or ERROR for each name sent over the network. If the number of End-device modules or their names change, you could use the ND – Node Discovery command to determine the modules you now have in a network.

Step 10. A Coordinator module can provide information about the association process and status when sent an AI – Association Indication command. See the Digi manual "XBee/XBee-PRO RF Modules" for complete information about all 20 responses an AI command could produce.

Within the X-CTU Terminal window, type +++ and wait for the OK reply. Then, type: ATAI[Enter]. The result should display only a zero, 0, to indicate a successful completion of the association between the Coordinator and End-device modules.

Other responses include error messages and status information about locating a PAN, timing out of a scan for unused frequencies, and so on. When you write software for a non-beacon network of several End-device modules, you should include steps that issue an AI command and test for a returned value of 0 before the program proceeds. If the returned value does not equal zero, your code should indicate the type of error or condition.

Step 11. Again issue an ATND command to the Coordinator module from the X-CTU Terminal to ensure all your End-device modules remain in your network. If they do not, remove power from the End-device modules, wait a few seconds and reapply power. The modules should join the Coordinator's network and their Associated LED should flash.

Turn off power to one of your End-device modules. Wait a few seconds and use the X-CTU Terminal window to send an ATND node-discover command to the Coordinator. You should see a response only from the powered End-device module or modules.

Turn power on to the unpowered End-device module and watch its LED. How long does it take for this End-device module to re-associate with the Coordinator? I found it took four to five seconds with two other End-device modules in my network. So if an End-device XBee module loses power and has power restored, the XBee module quickly associates with the network Coordinator. An End-device module in a network with many more modules might take a bit longer to re-associate. Use the ATND command again to ensure the Coordinator can discover all your End-device modules.

Step 12. Disconnect the Coordinator and USB-to-XBee adapter from the USB cable to turn off the Coordinator. Did you notice any change at the End-device LEDs? The LEDs will continue to flash as if they still belonged to the Coordinator's network. The LEDs on my three End-device modules still flashed 30 minutes after I disconnected the Coordinator. As a result, I recommend using the Associated LED for testing, rather than rely on it for a true indication of network association.

Re-connect the Coordinator and USB-to-XBee adapter to the USB cable attached to your PC. Now you will use the X-CTU Terminal window to send the Coordinator an ATND command. The X-CTU Terminal software has a quirk or bug so it will not communicate with the Coordinator until you first go to the PC Settings window and click on Test/Query. So do this now and wait for message "Communication with modem..OK." Then click on OK and switch to the Terminal window.

Type +++ , wait for the OK reply, and type ATND[Enter]. What do you observe? My Coordinator module simply replied with a 0x0D byte for a "carriage return." When you have powered End-device modules in a network and the Coordinator loses power, it will not automatically re-associate with the End-device modules. To force a re-association, you send the Coordinator an FR – Force Reset command.

In the Terminal window, type +++ and wait for the OK reply. Then type ATFR[Enter]. You should see an "OK" in the Terminal window along with two packets in the hex window:

`7E 00 02 8A 01 74` and `7E 00 02 8A 06 6F`

The 0x8A byte indicates a modem-status response. In the first packet, the 0x01 byte signifies a watch-dog timer reset and in the second packet, the 0x06 indicates a "coordinator started" condition. Now if you type +++ and then the ATND[Enter] sequence, you will find your End devices have re-associated with the Coordinator.

You can reset an XBee module by placing its Reset pin at ground for at least 200 nanoseconds. When I grounded the Reset pin my reset Coordinator replied:

`7E 00 02 8A 00 75` and `7E 00 02 8A 06 6F`

In the first packet, the 0x8A and 0x00 bytes signify a hardware reset and in the second packet, the 0x8A and 0x06 bytes indicate a "coordinator started" condition. But after waiting for at least 15 seconds after grounding the Reset pin and seeing the response above, sending the Coordinator an ATND command yielded mixed results. Most of the time the reset Coordinator replied with 0x0D rather than the expected information from the End-device modules.

A Coordinator will likely lose power at some time, so I recommend an attached microcontroller issue an ATFR command after it resets from power loss and before your code tries to communicate with the Coordinator. It cannot hurt to pulse the XBee module Reset pin, but send the ATFR command, too.

If you wish, you can send the Force Reset command in an API command packet rather than as an AT command:

```
07 00 04 08 52 4E 44 0D
```

It shouldn't take more than a few seconds for your End-device modules to re-associate with the reset Coordinator.

CONCLUSION

In this experiment you learned how to configure a Coordinator module and one or more End-device modules for use in a network. You saw how End-device modules can quickly associate with a Coordinator and how you can obtain information about the modules in a network. You also learned how to address an End-device module by name and how to check the status of the network association process. You saw the importance of using the Force Reset command to re-establish a network when a Coordinator Module temporarily loses power.

Digital and Analog Measurements in a Network

REQUIREMENTS

2 or 3 XBee modules
1 or 2 XBee adapter boards
2 Solderless breadboards
1 USB-to-XBee adapter
1 USB cable—type-A to mini-B
1 Microcontroller with a serial port (Arduino Uno or ARM mbed)
1 5-V-to-3.3-V logic-conversion circuit or module (Arduino Uno only)
1 3.3-V-to-5-V logic-conversion circuit or module (Arduino Uno only)
1 10-kohm potentiometer
1 10-kohm, 1/4 watt, 10% resistor (brown-black-orange)
1 4700-ohm, 1/4 watt, 10% resistor (yellow-violet-red)
1 1000-ohm, 1/4 watt, 10% resistor (brown-black-red)
3 330-ohm, 1/4 watt, 10% resistors (orange-orange-brown)
3 LEDs
Terminal-emulation software such as HyperTerminal for Windows (ARM mbed only)
Digi X-CTU software running on a Windows PC, with an open USB port

INTRODUCTION

In previous experiments you learned how to use AT commands or API packets to communicate with XBee modules and you learned how to associate modules in a small personal-area network (PAN). Those steps also showed you how to obtain information about End-device modules in your network. In this experiment, an MCU will issue a command to one or more End-device

195

modules in a network to obtain data from their analog and digital inputs. You performed similar steps with individual XBee modules in Experiment 14.

This experiment designates one XBee module as a Coordinator and one or more XBee modules as End devices. In my lab, I used the module named RCVR as the Coordinator and three modules named XMTR, PRO, and END as End devices. After you complete this experiment, you can go directly to Experiment 21 with the same equipment setup, but you may turn off power to your equipment. You can download the MCU code and XBee configuration files for this experiment at: http://www.elsevierdirect.com/companion. jsp?ISBN=9780123914040.

At this point, you should know how to use the X-CTU program without detailed instructions about what window to open, what control to click, and so on. You also should know how to set up and use a terminal-emulation program that can display information received from an MCU module. I'll provide details only when necessary.

Step 1. End Module or Modules: If you don't know the configuration of your XBee module, I recommend you restore it with the factory-default values. Within the X-CTU window, click on the Modem Configuration tab and in turn place each module in the USB-to-XBee adapter and click on Restore under the Modem Parameters and Firmware heading. You will find the Modem Configuration profile for the End-device modules in the file EX20_End.pro. If you load the configuration file into an X-Bee module, you must change the NI – Network Identifier information so you have a different "name" for each module.

Place one of your designated End-device XBee modules in the USB-to-XBee adapter and use the X-CTU software to Read its Modem Configuration information.

- Ensure you have a value of 0 for CE – Coordinator Enable and a value of 0x1FFE for SC – Scan Channels.
- For the A1 – End Device Association, select the setting 6 – 0110B, where the B stands for binary. This setting establishes the conditions: associate with a Coordinator on any channel, and attempt to associate with a Coordinator indefinitely.
- Move down to the last item under the Networking & Security heading: NI – Node Identifier and click on this label. Next, click on the Set button that appears to the right of this label and type in a name for the module. You can use as many as 20 characters, but I recommend you use only four or five.
- Look under the Serial Interfacing heading for the label, AP – API Enable, and click on it. Choose 1 – API ENABLED.
- In Table 20.1, write the name (NI) of each End-device module and its SH, and SL information. You will need this information later so you can compare it with responses from these modules and modify settings in the MCU software.

Table 20.1 Information for Your End-Device Modules

	Module 1	Module 2	Module 3
NI – Node Identifier			
SH – Serial Number High			
SL – Serial Number Low			

Note: Modules 2 and 3 optional.

Double check your Modem Configuration settings with those shown here:

CE – Coordinator Enable	= 0
SC – Scan Channels	= 0x1FFE
NI – Node Identifier	= your choice of 3 or 4 characters
AP – API Enable	= 1 – API ENABLED
A1 – End Device Association	= 6 – 0110B
RN – Random Delay Slots	= 2

DO NOT click Write.

- Move down to the section labeled I/O Settings and change your End-device module settings to match the ones below. DO NOT change the settings for D7 or D5. DO NOT change any other settings.

D3 – DIO3 Configuration	2 – ADC	Analog-to-digital converter
D2 – DIO2 Configuration	3 – DI	Digital input
D1 – DIO1 Configuration	2 – ADC	Analog-to-digital converter
D0 – DIO0 Configuration	3 – DI	Digital input

Double check your DIO settings.

- Finally, click on Write to save this configuration in the attached module. This step enables the API interface on your modules, gives it a name you can recognize, sets it as an End device, and enables four I/O pins as inputs. Set aside this End-device module to keep it separate from the Coordinator XBee module configured in the next steps.

Repeat Step 1 for each XBee module you will use as an End device and set aside these modules.

Step 2. Coordinator Module: Place the XBee module designated as your Coordinator in the USB-to-XBee adapter and use the X-CTU software to Read its Modem Configuration information. You will find the Modem Configuration profile for the Coordinator modules in the file EX20_Coord.pro.

- Under the heading Networking & Security, go to the MY – 16-Bit Source Address. Type in as many as four hexadecimal values.
- Ensure you have a value of 0x1FFE for SC – Scan Channels.
- In the line labeled CE – Coordinator Enable, select 1 – COORDINATOR which lets the Coordinator find an unused wireless channel, and lets any End device associate with it.

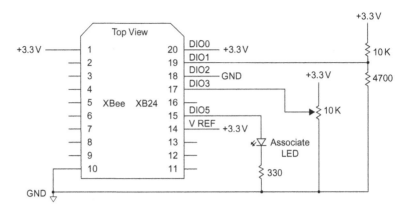

FIGURE 20.1 Circuitry needed to provide an End-device module with two digital and two analog inputs.

- For the A2 – Coordinator Association setting, select 6 – 110B, where the B stands for binary. This setting establishes the conditions: find an unused wireless channel and let any End device associate with this Coordinator.
- Move down to the last item under the Networking & Security heading: NI – Node Identifier and click on this label. Next, click on the Set button that appears to the right of this label and type in a name for the module. You can use as many as 20 characters, but I recommend you use four or five.
- Look under the Serial Interfacing heading for the label, AP – API Enable, and click on it. Choose 1 – API ENABLED.
- Finally, click on Write to save this configuration in the attached module. This step enables the API interface on your modules, gives it a name you can recognize, and sets it as an End device. (I gave my Coordinator the name RCVR.)

Leave the Coordinator module in the USB-to-XBee adapter connected to your PC.

Step 3. Place one of your End-device modules in an XBee adapter on your solderless breadboard and add the components shown in Figure 20.1. If the module has any other components or wires attached to it, please remove them now. The two digital inputs and two analog inputs provide information this End-device module will transmit to the Coordinator module.

Step 4. If you have configured more than one End-device module, ensure you have turned off power to your breadboard and place the second End-device module in an XBee adapter and insert it into your solderless bread-board. Connect the LED-and-resistor circuit at this module's DIO5 – Associate pin (pin 15) as shown in Figure 20.2. Then add the other connections to ground and +3.3 volts as shown in the same figure. If this XBee adapter has any other components or wires connected to it, remove them now. If you have more than two End-device modules, I recommend you place additional End-device modules in separate solderless breadboards.

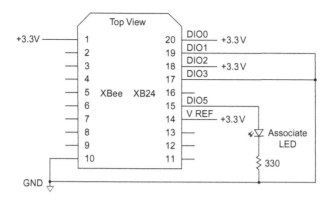

FIGURE 20.2 Minimum added circuitry for a second, or other End-device modules.

Step 5. After you have set up the circuitry shown in Figure 20.1—and in Figure 20.2 for additional End-device modules—turn on power to these modules. The End-device XBee module or modules should associate with the Coordinator attached to your PC, and the LED at each End-device module should flash to indicate proper network association. Use the X-CTU Terminal window to send an ATND command to the Coordinator to ensure it can discover all End-device modules, which should respond with their addresses. Remember these hexadecimal addresses do not include leading zeros.

Step 6. The set of XBee commands includes IS – Force Sample, which causes an XBee module to immediately reply with information from all enabled digital I/O lines and 10-bit analog-measurement values. So I thought use of the IS command in a broadcast API packet would force all End-device modules to immediately respond with their analog and digital information from their I/O pins. But based on experiments I performed, and duplicated by people at Digi International, the IS command does not behave this way. Thus we must address individual End-device modules with their serial number, or 64-bit (8-byte) address.

Use the X-CTU Terminal to send an API command packet for the IS – Force Sample command to one of your End-device modules identified by its 64-bit address. My End-device module named XMTR has the serial number SH = 0x0013A200 and SL = 0x4049E0EC, so I used the following packet. The underlined bytes represent the address of the XMTR module I want to execute the IS command:

7E 00 0F 17 52 <u>00 13 A2 00 40 49 E0 EC</u> FF FE 02 49 53 FD

I inserted the address as shown and calculated the checksum to end the packet. The following information from past experiments will refresh your knowledge of the API command packet format:

7E   = start byte

000F   = 15 bytes in the message

17	= Remote AT Command Request byte
52	= frame identifier value (all experiments use this value)
0013A2004049E0EC	= 64-bit End-device address
FFFE	= value needed to force 64-bit addressing above
02	= value that causes immediate actions
49	= hex code for letter I
53	= hex code for letter S
FD	= checksum

The response packet from my End-device module appeared in the following familiar format:

```
7E 00 18 97 52 00 13 A2 00 40 49 E0 EC FF FE 49 53 00 01 14 05
00 01 01 47 03 FF 0E
```

This packet breaks down as follows:

7E	= start byte
0018	= message length (24 bytes)
97	= packet type (Remote Command Response)
52	= frame-identifier byte (52 for all experiments)
0013A2004049E0EC	= 64-bit address of responding module
FFFE	= 16-bit value set for 64-bit addressing
49	= hex code for letter I
53	= hex code for numeral S
00	= status byte
01	= number of samples of digital and analog data
14	= first Active-Signal Byte, AD3 and AD1 active
05	= second Active-Signal Byte, DIO2 and DIO0 active
00	= first Digital-Data Byte
01	= second Digital-Data Byte, DIO2 = 0, DIO0 = 1
0147	= hex value from AD1 ADC
03FF	= hex value from AD3 ADC
0E	= checksum for this message

Step 7. You do not want to manually configure a network and obtain 64-bit addresses from each End-device module before someone installs them. That would take a lot of work and someone might inadvertently include errors in the list of hexadecimal addresses. You can automate the process. In this experiment you will see how to automate collection of the analog and digital information. In the next experiment you will learn how to discover modules when you don't know how many exist in a network.

In Experiment 18 you learned how a microcontroller (MCU) attached to a Coordinator can transmit an ND – Node Discovery command and obtain 64-bit module addresses and node-identifier information from all associated End devices. Once an MCU attached to a Coordinator has the 64-bit addresses for End-device modules it can send commands to specific modules. This experiment will use software to send an API command packet for the IS command. The software will include a standard packet for the IS command, into which it inserts the address for an End-device module. The length of the message will remain constant because all End-device modules have 64-bit, or 8-byte, addresses. Thus, the template for the IS API command packet looks like:

```
7E 00 0F 17 52 __ __ __ __ __ __ __ __ FF FE 02 49 53 ??
```

The blank spaces leave openings for the eight bytes of an End-device module address, with the most-significant address byte first (left) and the least-significant byte last (right). The template cannot include the checksum, because it will change based on the address values. But the software will calculate the checksum as the MCU transmits the packet bytes. Then it will append the checksum byte at the end of the packet. Remember, the checksum does not add the packet-start byte 0x7E or the following two bytes that indicate message length. And, obviously, it cannot add itself.

The people at Digi International told me it's unlikely the SH, 0x0013A200, portion of XBee module addresses (or identification number) will ever change because the remaining four address bytes provide for as many as 4.3 billion addresses, which cover many XBee modules. So you can simplify the template and include the four SH address bytes right from the start. Then the software need insert only the four SL address bytes and the checksum. The program in this experiment uses the simplified format with the four SL address bytes shown as blank spaces below:

```
7E 00 0F 17 52 00 13 A2 00 __ __ __ __ FF FE 02 49 53 ??
```

Step 8. To keep the software simple, you will enter the SH, SL and NI information for each End-device module you plan to communicate with into an array of bytes rather than use an ND API command packet to gather that information from End-device modules. This simulation approach simplifies testing the code. Programmers take a similar approach and simulate information that lets them test programs before they connect MCUs to real hardware. It costs less to test furnace-control software with data that simulates furnace behavior than to connect the MCU to the furnace and have a software bug blow it up!

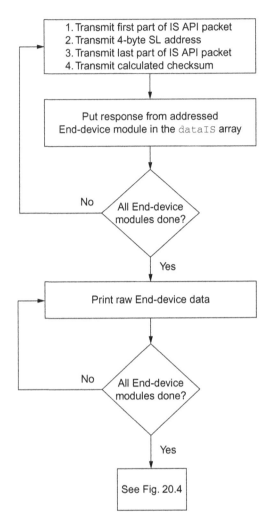

FIGURE 20.3 The flow chart for the software in this experiment shows how a program requests analog and digital data from each End-device module in your personal-area network. The program listing provides details.

The flow charts in Figures 20.3 and 20.4 show how the program works, but they do not include details such as all the declarations of variables and arrays or the SerialInput routine. You can read the comments in the complete listing for the Arduino Uno or ARM mbed boards to better understand that information. The software first transmits an IS API command packet to the first End-device identified in the dataND array entry. After it transmits the packet, the MCU waits for a response and saves it in another array named dataIS. As soon as the software has received responses from all identified End-device modules it prints the raw End-device data as hexadecimal values and checks that data for any errors. If it detects an error, it prints an error message and

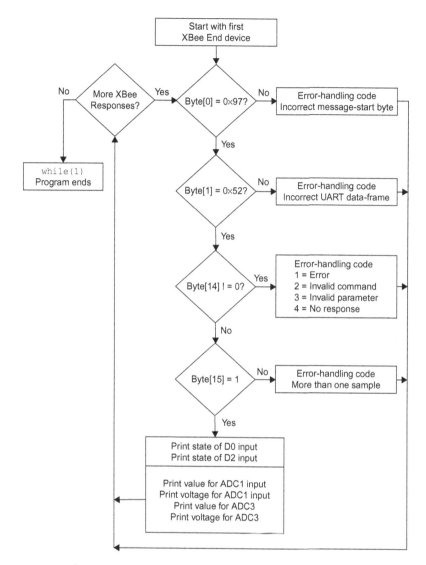

FIGURE 20.4 This flow chart shows the error-checking steps performed in the software for this experiment.

goes on to the next module's data. The flow chart in Figure 20.4 shows the error-checking steps.

If data for an End device comes through error free, which it almost always does, the program goes through each End-device module's data and extracts the information for the digital D0 and D2 inputs and the two analog inputs on each module. The formatted results look like:

XBee Module: XMTR

D0 = Logic 0

D2 = Logic 1

A1 Value = 335
A1 Voltage = 1.08
A3 Value = 810
A3 Voltage = 2.61

Important Note: If you send an IS (Force Sample) command to a module with no active I/O pins, that module indicates an error in its status byte in the reply to the network Coordinator.

The code for each MCU addresses specific elements in the `dataIS` array to obtain information received from each End-device module. Thus, the location `dataIS[x][15]` points to the number of I/O data samples in the packet and `dataIS[x][14]` refers to the status byte received from a module. The arrangement of bytes conforms to the structure shown prior to Step 7 in this experiment.

Step 8. Now you will add a microcontroller to your network to control the Coordinator. The following sections use an ARM mbed module or an Arduino Uno module.

USING AN ARM MBED MODULE

Turn off power to the breadboard and leave the End-device module with the resistors and potentiometer attached to it in the breadboard (see Figure 20.1). Remove the Coordinator module from the USB-to-XBee adapter, and set it aside.

If you have a second End-device module in your breadboard, remove it, and set it aside. If you do not have a second adapter socket in your breadboard, insert one and add the LED and resistor shown in Figure 20.5. Make the power (pin 1) and ground (pin 10) connections to the second adapter. If this adapter has any other connections, please remove them now. Connect the ARM mbed module to this adapter and insert the Coordinator module.

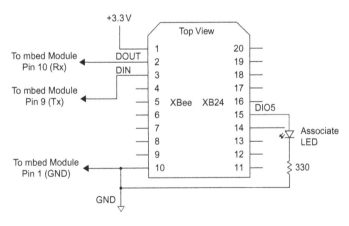

FIGURE 20.5 Circuit for the XBee module used in a breadboard as the Coordinator with an ARM mbed module.

If you have set aside a second End-device module, you can place it in another breadboard and add the connections shown earlier in Figure 20.2. You also could place an additional End-device module in the USB-to-XBee adapter. This XBee module would obtain power through the USB cable, but it will not communicate with the PC.

The photograph in Figure 20.6 shows a typical setup of a Coordinator and an End-device module. In my lab, I had additional End-device modules powered by two D-size dry cells on tables several meters from the Coordinator.

The program used for the ARM mbed will print the raw data as received by the MCU from End-device modules followed by formatted data about the digital and analog signals. If you do not want to see the raw hexadecimal information, "comment out" or remove this section of code.

Important: The code for this experiment assumes you have three (3) End-device modules. If you have a different number, change the boldface value in the statement below to indicate the proper number of End-device modules you will use in this experiment:

```
//Define number of modules here as a constant
const int numb_of_modules = 3;
```

Important: The code also assumed a 3.3-volt reference for the analog-to-digital converter in the End-device module wires as shown in Figure 20.1. If you use a different external voltage reference, say 2.5 volts, change the Vref declaration in the code to your reference voltage:

```
//define ADC reference-voltage input
float Vref = 3.3;
```

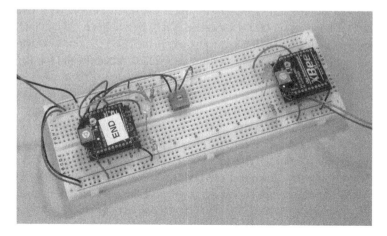

FIGURE 20.6 This image shows the arrangement of the End-device and Coordinator modules in a solderless breadboard.

In the array declaration that follows you enter the SH and SL address information (underlined) and the NI bytes (highlighted) for each End-device module. For a long NI "name," add bytes as needed for additional ASCII character values in hexadecimal format. Just ensure you end the group of NI bytes with a null, 0x00. My NI names in the `dataND` declaration correspond to XMTR, PRO, and END (see Table 20.1).

```
//Set up array with End-device 64-bit address starting at
dataND[ ][7]
//remember, dataND[ ][ ] array starts with dataND[0]
char dataND[3][dataNDlength] = {
{0x00, 0x00, 0x00, 0x00, 0x00, 0x00, 0x00,
0x00, 0x13, 0xA2, 0x00, 0x40, 0x49, 0xE0, 0xEC, 0x00, 0x58,
0x4D, 0x54, 0x52, 0x00},

{0x00, 0x00, 0x00, 0x00, 0x00, 0x00, 0x00,
0x00, 0x13, 0xA2, 0x00, 0x40, 0x6A, 0xF5, 0xAB, 0x00, 0x50,
0x52, 0x4F, 0x00, 0x00},

{0x00, 0x00, 0x00, 0x00, 0x00, 0x00, 0x00,
0x00, 0x13, 0xA2, 0x00, 0x40, 0x49, 0xE1, 0xE6, 0x00, 0x45,
0x4E, 0x44, 0x00, 0x00}
};
```

You will see the messages sent by the ARM mbed module on a terminal-emulator program set for 9600 bits/second, 8 data bits, no parity, 1 stop bit (8N1), and no flow control. You will not see anything in the X-CTU terminal window.

You can load the code shown in Program EX20_mbed_IS and run it. When you compile the code you will get warning messages "Expression has no effect" and "Expected a statement," but you can ignore them. The program will compile properly and you can download it to your ARM mbed module.

After the LED on the ARM mbed board stops flashing, start your terminal emulator program and set it for 9600 bits/second, 8 data bits, no parity, 1 stop bit, and no flow control. The terminal emulator uses the same USB cable the compiler used to download the code. You can use the Windows Device Manager to determine which virtual serial port the ARM mbed connects to. Change your view so you can see the terminal emulator window.

Remember to turn on power to your Coordinator and End-device modules in the breadboard. Press the reset button on the ARM mbed module to start the EX20_mbed_IS program.

Program EX20_mbed_IS

```
/*
 * EX20 ARM mbed Module   Rev. F
 * Program EX20_mbed_IS
 * Send IS API command packet to Coordinator for each End-device
 * module identified in dataND array. Save information in dataIS array,
 * print raw results as hex data, and print formatted analog and
 * digital data on PC terminal emulator.
 * Output to LED on pin 13 indicates errors.
 * Jon Titus 09-04-2011 @ 1120H
 */

#include "mbed.h"

//*************** Declarations & Assignments ***************

//Create start-of-API packet for IS command with this array of
//bytes.
//Byte count 0x0F does not include first 3 bytes or the checksum
//byte.
int packetIS_start[] = {0x7E, 0x00, 0x0F, 0x17, 0x52, 0x00, 0x13,
                        0xA2, 0x00};
int packetIS_start_len = 9;

//Create end-of-API packet for IS command
int packetIS_end[] = {0xFF, 0xFE, 0x02, 0x49, 0x53};
int packetIS_end_len = 5;

//Define number of modules here as a constant
const int numb_of_modules = 3;

//Define number of bytes in arrays as constants
const int dataISlength = 40;
const int dataNDlength = 40;

//Set up array with End-device 64-bit address starting at
//dataND[x][7]
//remember, dataND[] array starts with dataND[x][0]
//array definitions require a constant value
char dataND[numb_of_modules][dataNDlength] = {
{0x00, 0x00, 0x00, 0x00, 0x00, 0x00, 0x00,
0x00, 0x13, 0xA2, 0x00, 0x40, 0x49, 0xE0,
0xEC, 0x00, 0x58, 0x4D, 0x54, 0x52, 0x00},
{0x00, 0x00, 0x00, 0x00, 0x00, 0x00, 0x00,
 0x00, 0x13, 0xA2, 0x00, 0x40, 0x6A, 0xF5,
 0xAB, 0x00, 0x50, 0x52, 0x4F, 0x00, 0x00},
{0x00, 0x00, 0x00, 0x00, 0x00, 0x00, 0x00,
 0x00, 0x13, 0xA2, 0x00, 0x40, 0x49, 0xE1,
 0xE6, 0x00, 0x45, 0x4E, 0x44, 0x00, 0x00}
};

//Create an array for the response from each End-device module
char dataIS[numb_of_modules][dataISlength];

//Define bit masks for expected digital bits at DIO0 and DIO2 pins
int D0_mask = 0x01;        // 00000001
int D2_mask = 0x04;        // 00000100

//Define ADC reference-voltage input, change as needed.
float Vref = 3.3;
```

```
//Define serial I/O ports: "Sport" for Coordinator, "pc" for
//terminal emulator at your PC
Serial Sport(p9, p10);
Serial pc(USBTX, USBRX);

//Define digital-output pin for optional "error" LED
DigitalOut  LedPin(p13);

//Define additional temporary variables
int            XBee_numb;
int            testdata;
int            MSbits, LSbits;
unsigned int   bytecount_hi;
unsigned int   bytecount_lo;
unsigned int   bytecount;
unsigned int   counter;
unsigned int   chksum;
unsigned int   chksum_temp;
int            analog_temp;
float          analog_voltage;
int            NIcounter;

//*************** Serial-Input Routine ***************
//Function SerialInput reads UART after it has new data
int SerialInput()
{
    while (Sport.readable() == 0)  //wait for UART to have new data
    {
    }
    return (Sport.getc());         //get new data, return
}

//*************** Main Program ***************
int main(void)
{

//Send IS command to each End-device module and get results
XBee_numb = 0;
for (XBee_numb; XBee_numb < numb_of_modules; XBee_numb++)
{
    //Transmit IS packet with 64-bit address
    chksum = 0;
    //Start with array counter = 0
    counter = 0;

    //Transmit the start-of-API packet bytes and calculate checksum
    //only on the message portion of the packet. Do not compute
    //checksum on first three bytes--0x7E and byte count
    while(counter < packetIS_start_len)
        {
        Sport.putc(packetIS_start[counter]);
        if (counter > 2)            //Start checksum only after
            {                       //first three packet bytes sent
            chksum = chksum + packetIS_start[counter];
            }
        counter++;
        }
```

```
    //Transmit 64-bit (8-byte) address from dataND[] array
    //start at byte 11 in the dataND[ ][ ] array.
    counter = 11;
    while (counter < 15 )
        {
        Sport.putc(dataND[XBee_numb][counter]);
        chksum = chksum + dataND[XBee_numb][counter];
        counter++;
        }


//Transmit the end-of-API packet, reset counter for
    //packetIS_end array
    counter = 0;
    while (counter < packetIS_end_len)
        {
        Sport.putc(packetIS_end[counter]);
        chksum = chksum + packetIS_end[counter];
        counter++;
        }

    //Calculate checksum and send it
    //AND operator (&) saves only eight least-significant
    //bits for the subtraction
    Sport.putc(0xFF - (chksum & 0xFF));

    //Get responses from End-device modules via mbed serial port
    testdata = SerialInput();
        if (testdata == 0x7E);        //first serial byte == 0x7E?
        {                             //yes
        bytecount_hi = SerialInput();
        bytecount_lo = SerialInput();
        bytecount = (bytecount_hi * 256) + bytecount_lo;
                                      //get # of bytes in msg
        chksum_temp = 0;              //clear checksum value
        for (counter = 0; counter < bytecount; counter++)
                                      //put XBee data in array
            {
            dataIS[XBee_numb][counter] = SerialInput();
            chksum_temp = chksum_temp + dataIS[XBee_numb][counter];
            }
        chksum = SerialInput();       //get checksum--last byte
        if (chksum != (0xFF - (chksum_temp & 0xFF)))
                                      //do checksums match?
            {
            while(1)                  //NO MATCH, flash LED forever
                {                     //Error handling code could go
                LedPin = 0;           //here instead. For now
                wait(1);              //flash an LED
                LedPin = 1;
                wait(0.5);
                }
            }
        }                             //YES MATCH, do it for next module
    }


                                      //OK, got data for all XBee modules
wait(0.5);                            //half-second delay
```

```
//Routine to print data in each XBee array as hex characters
//Data goes to PC terminal emulator
XBee_numb = 0;
//print raw hexadecimal data in this loop, from here...
for (XBee_numb; XBee_numb < numb_of_modules; XBee_numb++)
    {
    pc.printf("\n\r");
        for (counter = 0; counter < dataISlength; counter++)
            {
            pc.printf("%02X  ", dataIS[XBee_numb][counter]);
            }
    pc.printf("\n\r");            //go to a new line in text window
    }
//to here.

//Process analog and digital data from each End-device module.
//First, check for transmission errors

for (XBee_numb = 0; XBee_numb < numb_of_modules; XBee_numb++)
    {
    if (dataIS[XBee_numb][0] != 0x97)
        {
        //error routine here--wrong response to remote API command
        pc.printf("Incorrect API identifier for module
        # %d \n\r\n\r", XBee_numb);
        goto error_exit;
        }

    if (dataIS[XBee_numb][1] != 0x52)
        {
        //error routine here--incorrect UART frame ID #, break out
        pc.printf("Incorrect UART frame API identifier for module
        # %d \n\r\n\r", XBee_numb);
        goto error_exit;
        }

    if (dataIS[XBee_numb][14] != 0)
        {
        pc.printf("Error %d for module
        #: %d\n\r\n\r",dataIS[XBee_numb][14], XBee_numb);
        goto error_exit;
        }

    if (dataIS[XBee_numb][15] != 1)
        {
        //handle sample error here, expected only 1 sample
        pc.printf("Incorrect sample number for module
        # %d \n\r\n\r", XBee_numb);
        goto error_exit;
        }

    //No errors, go ahead and process analog and digital data here
    //NIcounter points to start of Node Identifier name
    //in dataIS array, saved for each End-device module.
    //Print "XBee Module:" followed by NI name
    //Start with first byte of NI data in dataIS array
    NIcounter = 16;
    pc.printf("XBee Module: ");
```

```
while(dataND[XBee_numb][NIcounter] != 0x00)
                        //Print ASCII characters until null
    {
    pc.putc(dataND[XBee_numb][NIcounter]);
    NIcounter++;
    }
pc.printf("\n\r");

//Now print formatted digital information
//Test only bit D0 in digital-data bytes in dataIS array
if ((dataIS[XBee_numb][19] & D0_mask) > 0)
    {
    pc.printf("D0 = Logic 1\n\r");
    }
    else
        {
        pc.printf("D0 = Logic 0\n\r");
        }

//Test only bit D2 in digital data byte
if ((dataIS[XBee_numb][19] & D2_mask) > 0)
    {
    pc.printf("D2 = Logic 1\n\r");
    }
    else
        {
        pc.printf("D2 = Logic 0\n\r");
        }

        //Expect 2 analog values per End-device module
        //First input from AD1/DIO1 input pin
        analog_temp = dataIS[XBee_numb][20] * 256;
        analog_temp = dataIS[XBee_numb][21] + analog_temp;
        pc.printf("A1 Value:  %d\n\r", analog_temp);
        analog_voltage = analog_temp * Vref / 1024;
        pc.printf("A1 Voltage: %.2f\n\r", analog_voltage);

        //Second input from AD1/DIO1 input pin
        analog_temp = dataIS[XBee_numb][22] * 256;
        analog_temp = dataIS[XBee_numb][23] + analog_temp;
        pc.printf("A3 Value:  %d\n\r", analog_temp);
        analog_voltage = analog_temp * Vref / 1024;
        pc.printf("A3 Voltage:  %.2f\n\r", analog_voltage);

        //print two blank lines to separate information
        pc.printf("\n\r\n\r");

    error_exit:        //program comes here after printing an error
                       //message for a given module
    }
    while(1)           //end program in an infinite do-nothing
        {              //loop
        }

}                      //main() program ends here
```

After you run the program and see the formatted data, change the logic-inputs from ground to +3.3 volts, and vice versa on an End-device module. Adjust the potentiometer on your End-device module. You should see changes in the data.

Optional: Suppose weak batteries or a tripped circuit breaker causes one of the End-device modules to go "off line." How do you think the software will react? You can disconnect power to one of your end-device modules and run the software again. What happened?

The software still transmitted an IS API command packet to the "missing" End-device module, which cannot respond. The software gets stuck in a loop constantly waiting for that End-device module to respond. In the next experiment you will learn how to overcome that type of problem.

Optional: Turn off power to your modules and program one of your End-device modules so that you have disabled all of its AD/DIO pins, with the exception of D7 (1 – CTS FLOW CONTROL) and D5 (1 – ASSOCIATED INDICATOR). Put this End-device module back in its adapter and run the ARM mbed program again. The newly programmed module will cause an error because it has no I/O pins programmed. You should see an error message in the terminal-emulator window, but without the NI name. You could add program steps to identify the End-device module by name.

This optional step shows the importance of having the software configured to "look for" the same I/O configuration present in all End-device modules. You might use the analog and digital active-signal bytes to determine which digital and analog bits and bytes provide useful information, but that would take a lot more software.

Suppose you need two digital inputs on one End-device module, one digital and two analog inputs on another, and finally two digital and two analog inputs on a third. I recommend you simply configure all End-device modules for two digital inputs and two analog inputs. You can simply ignore the information you don't need. But all End-device modules have the same settings. This approach also simplifies replacing modules in the field.

ARM mbed Troubleshooting

If End-device modules do not associate with a Coordinator, recheck the configurations you saved in the modules when you set them in Steps 1 and 2. Check that you have power and ground properly connected to each XBee module in your breadboard.

Check the wiring of your breadboard circuits as shown in Figures 20.1, 20.2, and 20.5.

Ensure you have set the SH, SL, and NI information properly for each End-device module. Any error in the address bytes will cause the program to stall and you will see nothing happen. If you still have problems, recheck the first End-device module entry in the `dataND` array. Set the `numb_of_modules = 1` and try the program again. If you succeed, check the information for the second module and increase the `numb_of_modules` value to 2.

If the LED at your End-device or Coordinator module does not turn on, ensure you have it plugged in correctly. Try reversing the LED leads. They are sensitive to the polarity of current flow.

Check the configuration of the your terminal-emulator software to ensure you have 8 data bits, 1 stop bit, no parity, and no flow control for 9600 bit/second transmissions. Do you have the emulator set for the proper COM port? Have you told the emulator to connect to the ARM mbed computer via a COM port? Remember, you want the emulator to communicate with the ARM mbed module and not to any XBee module attached to the USB-to-XBee adapter. Although an End-device module plugged into the USB-to-XBee adapter receives only power from the adapter, your PC will still "see" the adapter as a COM port.

You have reached the end of this experiment for the ARM mbed module. If you plan to go immediately to Experiment 21, leave your hardware set up, but you may turn off power to your equipment.

USING AN ARDUINO UNO MODULE

Turn off power to the breadboard and leave the End-device module with the resistors and potentiometer attached to it in the breadboard (see Figure 20.1). Remove the Coordinator module from the USB-to-XBee adapter, and set it aside.

If you have a second End-device module in your breadboard, remove it, and set it aside. If you do not have a second adapter socket in your breadboard, insert one and add the LED and resistor shown in Figure 20.7. Make the power (pin 1) and ground (pin 10) connections, and connect the logic-level-conversion circuits and the Arduino Uno module as shown. If you have other connections to this adapter, please remove them and then insert the Coordinator module.

If you have set aside a second End-device module, you can place it in another breadboard and add the connections shown earlier in Figure 20.2.

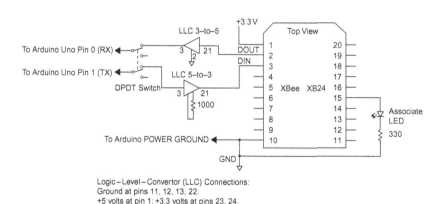

FIGURE 20.7 Circuit for the XBee module used as the Coordinator with an Arduino Uno module.

You also could insert an End-device module in the USB-to-XBee adapter. This XBee module would obtain power through the USB cable, but it will not communicate with the PC.

Pay attention to the notes below and in Figure 20.7 that explain additional connections not shown in Figure 20.7 for the sake of clarity:

- On LLC 5-to-3, connect a 1000-ohm resistor (brown-black-red) between pins 1 and 2.
- On LLC 3-to-5, connect pin 2 to ground.
- On both logic-level-converter devices, connect pins 11, 12, 13, and 22 to ground.
- On both logic-level-converter devices, connect pin 1 to +5 volts.
- On both logic-level-converter devices, connect pins 23 and 24 to +3.3 volts.

The image in Figure 20.8 shows the arrangement of an End-device module, a Coordinator module, and the logic-level-conversion circuits. In my lab, I had additional End-device modules powered by two D-size dry cells on tables several meters from the Coordinator.

The program used for the Arduino Uno will print the raw data as received by the MCU from End-device modules followed by formatted data about the digital and analog signals. If you do not want to see the hexadecimal information, "comment out" or remove this section of code.

Important: The code for this experiments assumes you have three (3) End-device modules. If you have a different number, change the boldface value

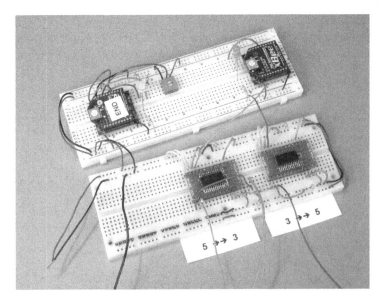

FIGURE 20.8 End-device module (left), Coordinator module (right), and logic-level-conversion circuits (bottom) in solderless breadboards.

in the statement below to indicate the proper number of End-device modules you will use in this experiment:

```
//Define number of modules here as a constant
const int numb_of_modules = 3;
```

Important: The code also assumed a 3.3-volt reference for the analog-to-digital converter in the End-device module wires as shown in Figure 20.1. If you use a different external voltage reference, say 2.5 volts, change the Vref declaration in the code to your reference voltage:

```
//define ADC reference-voltage input
float Vref = 3.3;
```

In the array declaration that follows enter the SH and SL address information (underlined) and the NI bytes (highlighted) for each End-device module. For a long NI "name," add bytes as needed for additional ASCII character values in hexadecimal format. Just ensure you end the group of NI bytes with a null, 0x00. My NI names in the dataND declaration correspond to XMTR, PRO, and END (see Table 20.1).

```
//Set up array with End-device 64-bit address starting at
dataND[ ][7]
//remember, dataND[ ][ ] array starts with dataND[0]
char dataND[3][dataNDlength] = {
{0x00, 0x00, 0x00, 0x00, 0x00, 0x00, 0x00,
0x00, 0x13, 0xA2, 0x00, 0x40, 0x49, 0xE0, 0xEC, 0x00, 0x58,
0x4D, 0x54, 0x52, 0x00},


{0x00, 0x00, 0x00, 0x00, 0x00, 0x00, 0x00,
0x00, 0x13, 0xA2, 0x00, 0x40, 0x6A, 0xF5, 0xAB, 0x00, 0x50,
0x52, 0x4F, 0x00, 0x00},


{0x00, 0x00, 0x00, 0x00, 0x00, 0x00, 0x00,
0x00, 0x13, 0xA2, 0x00, 0x40, 0x49, 0xE1, 0xE6, 0x00, 0x45,
0x4E, 0x44, 0x00, 0x00}
};
```

You will see the messages sent by the Arduino Uno in the Serial Monitor window in the Arduino compiler. You will not see anything in the X-CTU terminal window.

You can load the code shown in Program EX20_Uno_IS and run it. Remember to turn on power to your Coordinator and End-device modules in the breadboard. Ensure all circuits have a common ground. Press the reset button on the Arduino Uno module to start the EX20_Uno_IS program.

Important: As noted in other experiments, the Arduino Uno module shares its UART with an XBee module and the host PC USB connection, so maintaining the serial connection between a Uno module and an XBee module when downloading code can cause an electrical conflict. You MUST manually disconnect (or use a small switch to open) the two serial-port connections between an Arduino Uno module and any external serial device while the compiler downloads code to the Uno. I used a small DPDT toggle switch to disconnect the signals between the logic-level-conversion circuits and the Arduino Uno module during code downloads. Remember to reconnect these wires, or flip the switch, before you run the program.

Program EX20_Uno_IS

```
/*
 * EX20 Arduino Uno Module  Rev. B
 * Program EX20_Uno_IS
 * Send IS API command packet to Coordinator for each End-device
 * module identified in dataND array. Save information in dataIS
 * array, print raw results as hex data, and print formatted
 * analog and digital data in Serial Monitor window.
 * Output to LED on pin 13 indicates errors.
 * Jon Titus 09-04-2011 @ 1120H
 */

//*************** Declarations & Assignments ***************

//Create start-of-API packet for IS command with this array of
//bytes.
//Byte count 0x0F does not include first 3 bytes or the checksum
//byte.
byte packetIS_start[] = {0x7E, 0x00, 0x0F, 0x17, 0x52, 0x00, 0x13,
                         0xA2, 0x00};
int packetIS_start_len = 9;

//Create end-of-API packet for IS command
byte packetIS_end[] = {0xFF, 0xFE, 0x02, 0x49, 0x53};
int packetIS_end_len = 5;

//Define number of modules here as a constant
const int numb_of_modules = 3;

//Define number of bytes in arrays as constants
const int dataISlength = 40;
```

```
//Set up array with End-device 64-bit address starting at
//dataND[x][7]
//remember, dataND[] array starts with dataND[x][0]
//array definitions require a constant value
byte dataND[numb_of_modules][dataNDlength] = {
{0x00, 0x00, 0x00, 0x00, 0x00, 0x00, 0x00,
 0x00, 0x13, 0xA2, 0x00, 0x40, 0x49, 0xE0,
 0xEC, 0x00, 0x58, 0x4D, 0x54, 0x52, 0x00},

{0x00, 0x00, 0x00, 0x00, 0x00, 0x00, 0x00,
 0x00, 0x13, 0xA2, 0x00, 0x40, 0x6A, 0xF5,
 0xAB, 0x00, 0x50, 0x52, 0x4F, 0x00, 0x00},

{0x00, 0x00, 0x00, 0x00, 0x00, 0x00, 0x00,
 0x00, 0x13, 0xA2, 0x00, 0x40, 0x49, 0xE1,
 0xE6, 0x00, 0x45, 0x4E, 0x44, 0x00, 0x00}
};

//Create an array for the response from each End-device module
byte dataIS[numb_of_modules][dataISlength];

//Define bit masks for expected digital bits at DIO0 and DIO2 pins
byte D0_mask = 0x01;      // 00000001
byte D2_mask = 0x04;      // 00000100

//Define ADC reference-voltage input, change as needed.
float Vref = 3.3;

//Define digital-output pin for optional "error" LED
int    LedPin = 13;

//Define additional temporary variables
int           XBee_numb;
int           testdata;
int           MSbits, LSbits;
unsigned int  bytecount_hi;
unsigned int  bytecount_lo;
unsigned int  bytecount;
unsigned int  counter;
unsigned int  chksum;
unsigned int  chksum_temp;
int           analog_temp;
float         analog_voltage;
int           NIcounter;

//************** Serial-Input Routine **************
//Function SerialInput reads UART after it has new data
int SerialInput()
{
    while (Serial.available() == 0)
                              //wait for UART to have new data
    {
    }
    return (Serial.read());      //get new data, return
}

//************** Setup Operations **************
void setup()
```

```
{
  Serial.begin(9600);
  pinMode(LedPin, OUTPUT);
  digitalWrite(LedPin, HIGH);
}

//************** Main Loop **************
void loop()
{

//Send IS command to each End-device module and get results
XBee_numb = 0;
for (XBee_numb; XBee_numb < numb_of_modules; XBee_numb++)
{
    //Transmit IS packet with 64-bit address, clear checksum
    chksum = 0;
    //Start with array counter = 0
    counter = 0;

    //Transmit the start-of-API packet bytes and calculate
    //checksum only on the message portion of the packet. Do not
    //compute checksum on first three bytes--0x7E and byte count
    while(counter < packetIS_start_len)
        {
        Serial.write(packetIS_start[counter]);
        if (counter > 2)          //Start checksum only after
            {                     //first three packet bytes sent
            chksum = chksum + packetIS_start[counter];
            }

        counter++;
        }

    //Transmit 64-bit (8-byte) address from dataND[] array
    //start at byte 11 in the dataND[ ][ ] array.
    counter = 11;
    while (counter < 15 )
        {
        Serial.write(dataND[XBee_numb][counter]);
        chksum = chksum + dataND[XBee_numb][counter];
        counter++;
        }

//Transmit the end-of-API packet, reset counter for
    //packetIS_end array
    counter = 0;
    while (counter < packetIS_end_len)
        {
        Serial.write(packetIS_end[counter]);
        chksum = chksum + packetIS_end[counter];
        counter++;
        }

    //Calculate checksum and send it
    //AND operator (&) saves only eight least-significant
    //bits for the subtraction
    Serial.write(0xFF - (chksum & 0xFF));
```

```
    //Get responses from End-device modules via mbed serial port
    testdata = SerialInput();
        if (testdata == 0x7E);      //first serial byte == 0x7E?
        {                           //yes
        bytecount_hi = SerialInput();
        bytecount_lo = SerialInput();
        bytecount = (bytecount_hi * 256) + bytecount_lo;
                            //get # of bytes in msg
        chksum_temp = 0;        //clear checksum value
        for (counter = 0; counter < bytecount; counter++)
                            //put XBee data in array
            {
            dataIS[XBee_numb][counter] = SerialInput();
            chksum_temp = chksum_temp + dataIS[XBee_numb][counter];
            }
        chksum = SerialInput();   //get checksum--last byte
        if (chksum != (0xFF -(chksum_temp & 0xFF)))
                            //do checksums match?
            {
            while(1)            //NO MATCH, flash LED forever
                {               //Error handling code could go
                digitalWrite(LedPin, LOW);
                                //here instead. For now
                delay(1000);    //flash an LED
                digitalWrite(LedPin, HIGH);
                delay(500);
                }
            }
        }                       //YES MATCH, do it for next module

}

                            //OK, got data for all XBee modules
  delay(500);               //half-second delay

    //Routine to print data in each XBee array as hex characters
    //Data goes to PC terminal emulator
    XBee_numb = 0;
    //print raw hexadecimal data in this loop, from here...
    for (XBee_numb; XBee_numb < numb_of_modules; XBee_numb++)
        {
        Serial.print("\n\r");
            for (counter = 0; counter < dataISlength; counter++)
                {
                if (dataIS[XBee_numb][counter] < 0x10)
                  {
                  Serial.print("0");
                  }             //insert leading zero, if needed
                  Serial.print(dataIS[XBee_numb][counter], HEX);
                Serial.print(" ");
                }
        Serial.print("\n\r");   //go to a new line in text window
        }
    //to here.

    //Process analog and digital data from each End-device module.
    //First, check for transmission errors
```

```
for (XBee_numb = 0; XBee_numb < numb_of_modules; XBee_numb++)
    {
    if (dataIS[XBee_numb][0] != 0x97)
        {
        //error routine here--wrong response to remote API
        //command
        Serial.print("Incorrect API identifier for module # ");
        Serial.print(XBee_numb);
        Serial.print("\n\r\n\r");
        goto error_exit;
        }

    if (dataIS[XBee_numb][1] != 0x52)
        {
        //error routine here--incorrect UART frame ID #,
        //break out

Serial.print("Incorrect UART frame API identifier for module # ");
        Serial.print(XBee_numb);
        Serial.print("\n\r\n\r");
        goto error_exit;
        }

    if (dataIS[XBee_numb][14] != 0)
        {
        Serial.print("Error ");
        Serial.print(dataIS[XBee_numb][14]);
        Serial.print(" for module ");
        Serial.print(XBee_numb);
        Serial.print("\n\r\n\r");
        goto error_exit;
        }

    if (dataIS[XBee_numb][15] != 1)
        {
        //handle sample error here, expected only 1 sample
        Serial.print("Incorrect sample number for module # ");
        Serial.print(XBee_numb);
        Serial.print("\n\r\n\r");
        goto error_exit;
        }

    //No errors, go ahead and process analog and digital
    //data here
    //NIcounter points to start of Node Identifier name
    //in dataIS array, saved for each End-device module.
    //Print "XBee Module:" followed by NI name
    //Start with first byte of NI data in dataIS array
    NIcounter = 16;
    Serial.print("XBee Module: ");
    while(dataND[XBee_numb][NIcounter] != 0x00)
                            //Print ASCII characters until null
        {
        Serial.write(dataND[XBee_numb][NIcounter]);
        NIcounter++;
        }
    Serial.print("\n\r");

    //Now print formatted digital information
    //Test only bit D0 in digital-data bytes in dataIS array
```

```
if ((dataIS[XBee_numb][19] & D0_mask) > 0)
        {
        Serial.print("D0 = Logic 1\n\r");
        }
        else
            {
            Serial.print("D0 = Logic 0\n\r");
            }

    //Test only bit D2 in digital data byte
    if ((dataIS[XBee_numb][19] & D2_mask) > 0)
        {
        Serial.print("D2 = Logic 1\n\r");
        }
        else
            {
            Serial.print("D2 = Logic 0\n\r");
            }

    //Expect 2 analog values per End-device module
    //First input from AD1/DIO1 input pin
    analog_temp = dataIS[XBee_numb][20] * 256;
    analog_temp = dataIS[XBee_numb][21] + analog_temp;
    Serial.print("A1 Value:  ");
    Serial.print(analog_temp);
    Serial.print("\n\r");
    analog_voltage = analog_temp * Vref / 1024;
    Serial.print("A1 Voltage:  ");
    Serial.print(analog_voltage);
    Serial.print("\n\r");

    //Second input from AD1/DIO1 input pin
    analog_temp = dataIS[XBee_numb][22] * 256;
    analog_temp = dataIS[XBee_numb][23] + analog_temp;
    Serial.print("A3 Value:  ");
    Serial.print(analog_temp);
    Serial.print("\n\r");
    analog_voltage = analog_temp * Vref / 1024;
    Serial.print("A3 Voltage:  ");
    Serial.print(analog_voltage);
    Serial.print("\n\r");

//print two blank lines to separate information
        Serial.print("\n\r\n\r");

    error_exit:         //program comes here after printing an error
    while(0)  {  }      //do-nothing statement to satisfy compiler
//message for a given module
    }
    while(1)            //end program in an infinite do-nothing loop
    {
    }

}                       //main loop() ends here
```

After you run the program and see the formatted data, change the logic-inputs at an End-device module from ground to +3.3 volts, and vice versa. Adjust the potentiometer on your End-device module. You should see changes in the data when you again run the EX20_Uno_IS program. Just press the Arduino Uno reset pushbutton.

Optional: Suppose weak batteries or a tripped circuit breaker causes one of the End-device modules to go "off line." How do you think the software will react? You can disconnect power to one of your end-device modules and run the software again. What happened?

The software still transmitted an IS API command packet to the "missing" End-device module, which cannot respond. The software gets stuck in a loop constantly waiting for that End-device module to respond. In the next experiment you will learn how to overcome that type of problem.

Optional: Turn off power to your modules and reprogram one of your End-device modules so you have disabled all of its AD/DIO pins, with the exception of D7 (1 – CTS FLOW CONTROL) and D5 (1 – ASSOCIATED INDICATOR). Put this End-device module back in its adapter and run the Arduino Uno program again. The newly programmed module will cause an error because it has no I/O pins programmed.

This optional step shows the importance of having the software configured to "look for" the same I/O configuration present in the modules. You might use the analog and digital active-signal bytes to determine which digital and analog bits and bytes provide useful information, but that would take a lot more software.

Suppose you need two digital inputs on one End-device module, one digital and two analog inputs on another, and finally two digital and two analog inputs on a third. I recommend you simply configure all End-device modules for two digital inputs and two analog inputs. You can simply ignore the information you don't need. But all End-device modules have the same settings. This approach also simplifies replacing modules in the field.

Arduino Uno Troubleshooting

As noted earlier, you must disconnect connections from the Arduino Uno module RX←0 and TX→1 to the logic-level-conversion circuits when you compile and download a program. If you forget you might see an error message in the compiler text window.

If End-device modules do not associate with a Coordinator, recheck the configurations you saved in the modules when you set them in Steps 1 and 2. Check that you have power and ground properly connected to each XBee module in your breadboard. All circuits must have a common ground.

Check the wiring of your breadboard circuits as shown in Figures 20.1, 20.2, and 20.7. Recheck the connections to the logic-level-conversion circuits. It's easy to mix up the receiver and transmitter signals.

If the LED at your End-device module does not turn on, ensure you have it plugged in correctly. Try reversing the LED leads. They are sensitive to the polarity of current flow.

You have reached the end of this experiment for the Arduino Uno module. If you plan to go immediately to Experiment 21, leave your hardware set up, but you may turn off power to your equipment.

How to Handle an Unknown Number of XBee Modules

REQUIREMENTS

2 or 3 XBee modules
2 or 3 XBee adapter boards
1 or 2 Solderless breadboards
1 USB-to-XBee adapter
1 USB cable—type-A to mini-B
1 Microcontroller with a serial port (ARM mbed or Arduino Uno)
1 5-V-to-3.3-V logic-conversion circuit or module (for Arduino Uno only)
1 3.3-V-to-5-V logic-conversion circuit or module (for Arduino Uno only)
3 LEDs
3 330-ohm, 1/4 watt, 10% resistors (orange-orange-brown)
1 4700-ohm, 1/4 watt, 10% resistor (yellow-violet-red)
1 10-kohm, 1/4 watt, 10% resistor (brown-black-orange)
1 10-kohm potentiometer
Terminal-emulation software such as HyperTerminal for Windows (for ARM mbed)
Digi X-CTU software running on a Windows PC, with an open USB port

INTRODUCTION

In Experiments 18, 19, and 20 you learned how to associate XBee modules in a personal-area network (PAN), identify them, and obtain analog and digital information from them. Previously you preset microcontroller (MCU) software to work with one or more End-device modules. But you might not always know how many End-device modules a personal-area network (PAN) will have. An MCU that expects a preset number of modules could get "stuck" in an endless loop, waiting, waiting, and still waiting for information from modules

223

no longer in the PAN. Likewise, if new modules join the PAN, the MCU would miss receiving their data because it expected data from only x number of modules and it could not account for the new members of the PAN.

This situation causes a problem: How long should the MCU wait for responses from remote XBee modules before it goes on to other tasks? In this experiment you will learn how to use an MCU timer, interrupt, and interrupt-service routine (ISR). By using a timer, the MCU will wait sufficiently long for all modules to respond without knowing how many exist. Later you will learn about an alternate approach to this timing problem. The software in this experiment does not include a preset value for the number of modules in a PAN; the MCU will determine that number. The software examples will run on an Arduino Uno or on an ARM mbed module. The main differences center on the use of a timer specific to each MCU.

You can download the MCU code and XBee configuration files for this experiment at: http://www.elsevierdirect.com/companion.jsp?ISBN=9780123914040. If you have just completed Experiment 20 and have not changed your module configurations or circuits, skip ahead to Step 6. Otherwise, please continue with Step 1.

Step 1. End-Device Module or Modules: If you don't know the configuration of an XBee module, I recommend you restore it with the factory-default values. Within the X-CTU window, click on the Modem Configuration tab and in turn place each module in the USB-to-XBee adapter and click on Restore under the Modem Parameters and Firmware heading. You will find the Modem Configuration profile for the End-device modules in the file EX21_End.pro. If you load the configuration file into an X-Bee module, you must change the NI – Network Identifier information so you have a different "name" for each module.

Place one of your designated End-device XBee modules in the USB-to-XBee adapter and use the X-CTU software to Read its Modem Configuration information.

- Ensure you have a value of 0 for CE – Coordinator Enable and a value of 0x1FFE for SC – Scan Channels.
- For the A1 – End Device Association, select the setting 6 – 0110B, where the B stands for binary. This setting establishes the conditions: associate with a Coordinator on any channel, and attempt to associate with a Coordinator indefinitely.
- For the RN – Random Delay Slots, enter 2.
- Move down to the last item under the Networking & Security heading: NI – Node Identifier and click on this label. Next, click on the Set button that appears to the right of this label and type in a name for the module. You can use as many as 20 characters, but I recommend you use only four or five.
- Look under the Serial Interfacing heading for the label, AP – API Enable, and click on it. Choose 1 – API ENABLED.

Table 21.1 Information for Your End-Device Modules

	End-Device 1	End-Device 2	End-Device 3
NI – Node Identifier			
SH – Serial Number High			
SL – Serial Number Low			

Note: End-device modules 2 and 3 are optional.

- Write the name (NI) of each End-device module and its SH, and SL information in Table 21.1. You will need this information later so you can compare it with responses from these modules.

  Double check your Modem Configuration settings with those shown here:

  CE – Coordinator Enable   = 0

  SC – Scan Channels   = 0x1FFE

  NI – Node Identifier   = your choice of 3 or 4 characters

  AP – API Enable   = 1 – API ENABLED

  A1 – End Device Association   = 6 – 0110B

  RN – Random Delay Slots   = 2

  Do not click Write. (It's OK if you did, though. Just continue.)

- Move down to the section labeled I/O Settings and change your End-device module settings to match the ones below. DO NOT change the settings for D7 or D5 and DO NOT change any other settings.

  D3 – DIO3 Configuration   2 – ADC   Analog-to-digital converter

  D2 – DIO2 Configuration   3 – DI   Digital input

  D1 – DIO1 Configuration   2 – ADC   Analog-to-digital converter

  D0 – DIO0 Configuration   3 – DI   Digital input

  Double check your DIO settings.

- Finally, click on Write to save this configuration in the attached module. This step enables the API interface on your modules, gives it a name you can recognize, sets it as an End device, and enables four I/O pins as inputs. Set aside this End-device module to keep it separate from the Coordinator XBee module configured in the next steps.

Repeat Step 1 for each XBee module you will use as an End device and set aside these modules.

Step 2. Coordinator Module: Place the XBee module designated as your Coordinator in the USB-to-XBee adapter and use the X-CTU software to Read its Modem Configuration information. You will find the Modem Configuration profile for the Coordinator modules in the file EX21_Coord.pro.

- Ensure you have a value of 0x1FFE for SC – Scan Channels.
- In the line labeled CE – Coordinator Enable, select 1 – COORDINATOR which lets the module serve as a Coordinator in a PAN.

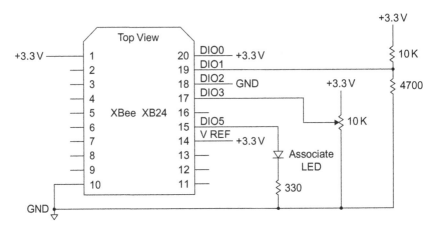

FIGURE 21.1 Circuitry needed to provide an End-device module with two digital and two analog inputs.

- For the A2 – Coordinator Association setting, select 6 – 110B, where the B stands for binary. This setting establishes the conditions: find an unused wireless channel and let any End device associate with this Coordinator.
- Move down to the last item under the Networking & Security heading: NI – Node Identifier and click on this label. Next, click on the Set button that appears to the right of this label and type in a name for the module. You can use as many as 20 characters, but I recommend you use four or five.
- Look under the Serial Interfacing heading for the label, AP – API Enable, and click on it. Choose 1 – API ENABLED.
- Finally, click on Write to save this configuration in the attached module. This step enables the API interface on your modules, gives it a name you can recognize, and sets it as an End device. (I gave my Coordinator the name RCVR.)

Leave the Coordinator module in the USB-to-XBee adapter connected to your PC.

Step 3. Place one of your End-device modules in an XBee adapter on your solderless breadboard and add the components shown in Figure 21.1. If the module has any other components or wires attached to it, please remove them now. The two digital inputs and two analog inputs provide information this End-device module will transmit to the Coordinator module.

This experiment can use an LED and 330-ohm resistor at pin 13 on the Arduino Uno or ARM mbed module as an optional status or error indicator. This LED circuit is not necessary to complete the experiment, but you may add it if you wish. (See Figure 18.6 for the LED-resistor circuit diagram.)

Step 4. If you have configured more than one End-device module, ensure you have turned off power to your breadboard and place the second End-device module in an XBee adapter and insert it into your

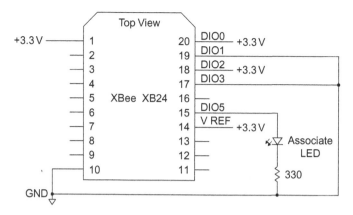

FIGURE 21.2 Minimum added circuitry for additional End-device modules.

solderless breadboard. Connect the LED-and-resistor circuit at this module's DIO5 – Associate pin (pin 15) as shown in Figure 21.2. Then add the other connections to ground and +3.3 volts as shown in the same figure. If this XBee adapter has any other components or wires connected to it, remove them now. If you have more than two End-device modules, I recommend you place additional End-device modules in separate solderless breadboards.

Step 5. After you have set up the circuitry shown in Figure 21.1—and in Figure 21.2 for additional End-device modules—turn on power to these modules. The End-device XBee module or modules should associate with the Coordinator attached to your PC, and the LED at each End-device module should flash to indicate proper network association. Use the X-CTU Terminal window to send an ATND command to the Coordinator to ensure it can discover all End-device modules, which should respond with their addresses. Remember the hexadecimal addresses do not include leading zeros. A response from *one* End-device module should appear in this form:

```
+++OK
ATND
FFFE
13A200
406AF5AB
51
PRO
```

Information will vary from that shown here due to different serial numbers and wireless signal strengths, but it always includes a complete response from each End device. If you do not get a response from a module, check its configuration settings. If an End device does not respond at all and its associate LED does not flash, check its electrical connects *and* its configuration information.

Step 6. Continue this experiment with either an ARM mbed or an Arduino Uno module after the following explanation of how the software operates.

MCU SOFTWARE

The software in this experiment will control the XBee Coordinator module and will receive SH and SL address information from all associated End-device modules. If the MCU does not receive a response from the Coordinator module within a set period, it "times out" and assumes all modules have replied. In this way it will not endlessly wait for nonexistent XBee modules to respond. Yet the MCU will wait long enough to obtain information from all associated End devices. For non-programmers, an abbreviated explanation, shown later in Table 21.2, will help clarify the overall serial-input operations.

PROGRAM DESCRIPTION

Flow charts in Figures 21.3a and 21.3b show how the program operates. After the MCU transmits the ND API command packet it begins the SerialInput routine, shown in an oval in Figure 21.3a and in detail in Figure 21.3b.

Instructions start an internal MCU timer and enable the timer's interrupt. The timer runs independently and doesn't interfere with other MCU operations, so in effect the timer runs in the "background." When the timer reaches the end of its preset period, established with the constant start_value, it causes an interrupt. The interrupt forces the MCU to leave the SerialInput routine and execute the instructions in the timer's associated interrupt-service routine (ISR), as diagrammed in Figure 21.4.

Because you cannot know when the timer will need attention, you cannot include statements to perform frequent timer operations in the normal program flow. But a timer that causes an interrupt requires immediate attention, so the MCU will stop what it is doing, mark its place so it can return to the main program, and branch to the timer ISR. After the MCU performs the operations in the ISR, it goes back to where it left off in the main program. Many other MCU devices use interrupts so the MCU will quickly respond to their operations.

The timer's simple ISR increments the timer_count value, resets the timer with the start_value, restarts the timer, and returns to the SerialInput routine. So timer interrupts occur at fixed intervals and each interrupt causes the timer_count value to increase by one.

As soon as the SerialInput routine receives information from the Coordinator module, it disables the timer interrupt. Then it returns the received serial byte to the main program for processing. The main program will continue to use the SerialInput routine to obtain additional bytes from the Coordinator module, but without using the timer or causing any interrupts. The timer and its ISR only get used in the program when the MCU waits for the first byte—the start byte, 0x7E—in a packet. That start byte suffices to let the MCU know the Coordinator has discovered another End-device module and has started to send its information to the MCU. After the MCU processes information from an End-device module, it turns the interrupt on again and waits for another start byte in a new packet.

(a)

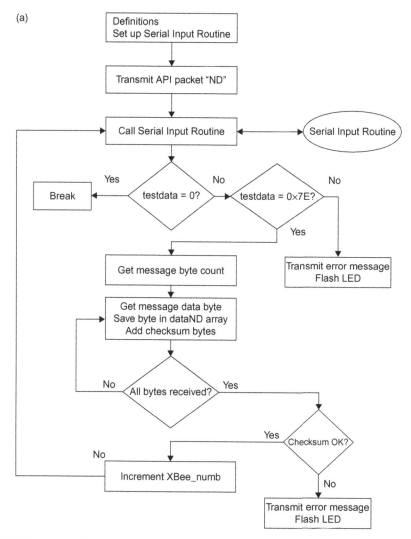

FIGURE 21.3 (a) Flow chart for code in the main routine that counts modules and uses a timer and an interrupt to break out of a loop waiting for nonexistent XBee modules. Note that the MCU can only exit this code via the Break shown on the left side of the diagram. (b) This flow chart shows the serial-input routine (top) that starts an 8-bit MCU up-timer with a preset value and enables a timer interrupt, and the interrupt-service routine (ISR) for the timer (bottom).

What happens when the SerialInput routine never receives another byte from the Coordinator module? The routine can now handle this condition because it continually tests the timer_count value to determine if it exceeds 255 (0xFF). Remember, the timer continues to periodically interrupt the serial-input routine and its ISR adds one to the timer_count value. When the timer has interrupted the serial-input routine more than 255

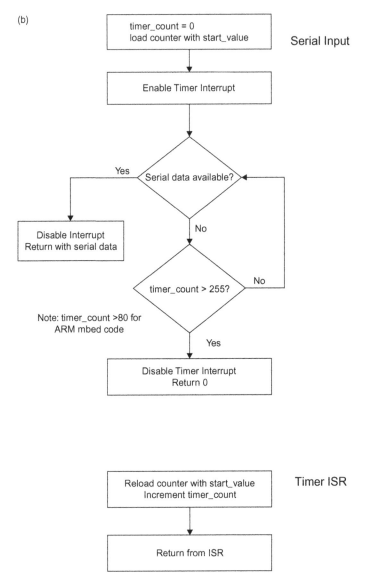

(b)

FIGURE 21.3 (Continued)

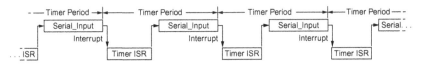

FIGURE 21.4 This timing diagram shows how the timer interrupts the Serial_Input function at fixed intervals. The MCU branches to the ISR, performs the ISR operations and then goes back to where the MCU left the Serial_Input function. (Not drawn to scale.)

times, the serial-input routine turns off the timer and returns to the main program with a `testdata` value of 0 (0x00). This value indicates the serial-input
routine has remained idle—no serial data received from the Coordinator—for
about eight seconds. (You can increase this period, if necessary.)

In essence, the result of the statement `if (testdata == 0)` indicates either
"Aha, there's a module out there; let's get all its data," or "I've waited for 8
seconds and received no additional response, so stop waiting and go back to
the main program."

Note that the software only checks the `testdata` value with the `if (tes-
tdata == 0)` statement when it waits for the *start byte* in a packet from the
Coordinator. If you try to test the `testdata` for a value of 0 when the software calls the `SerialInput` function at other places in the program, the program cannot distinguish between a 0 in the data from an XBee module and a 0
created when the MCU has waited for more than eight seconds.

Important: Remember that the software sends the ND API command packet
to the Coordinator, which queries all associated devices. The software DOES
NOT issue a *broadcast* ND API command packet. The latter will give you
unpredictable results. Again, the Coordinator received a *local* ND command.

Explanation for Non-programmers

Flow charts in Figures 21.3a and 21.3b show how the program operates, but
you can think of the operation of the first section of the program that involves a
timer and an interrupt as the steps shown in Table 21.2.

Table 21.2 Analogy for Program Steps

Step	Analogy	Program
1	You email several friends and ask them to come over to your lab bench, but you don't know how many will show up.	Transmit the ND API command packet.
2	You set your watch to beep every 10 seconds.	Set up timer.
3	You go to your office door and decide to wait five minutes, or 30 10-second periods.	`SerialInput` routine.
4	Each time your watch beeps, you make a mark.	Timer interrupt-service routine (ISR).
5	If a friend arrives, take him or her to your lab bench. Erase your timing marks and go back to the door.	Return from `SerialInput` routine with XBee data.
6	Have you made 30 marks yet? If not, go back to Step 4. If so, continue.	Within `SerialInput` routine.
7	Five minutes have elapsed since a friend arrived, so go to your lab bench and explain your latest project to the people there.	Return from `SerialInput` routine with `testdata = 0`.

Three key sections of C-language code:

- Set up a timer that will interrupt the processor. The timer has a long period.
- Include a short interrupt-service routine (ISR) that resets the timer and increments a count. Thus, whenever the timer reaches a set value, or overflows, the ISR increments the `timer_count` value.
- Include statements in the `SerialInput` routine that set the `timer_count` value to 0, load the MCU timer with a starting value, and turn on the timer's interrupt. When the timer exceeds a preset value, it interrupts the MCU and increments `timer_count`. When `timer_count` exceeds a preset value, the code aborts the `SerialInput` routine and returns a 0x00 value to the main program.

The code in this experiment sets up an array for a *maximum* of five (5) End-device modules when it defines the `dataND` array, but you can change the value of `numb_of_modules` depending upon the maximum number of modules you will allow in your PAN. You might include a few more than the maximum number of End-device modules expected. The definition:

```
dataND[numb_of_modules][dataNDlength]
```

simply assigns array space for data. Remember, array-element numbers start at zero; that is `dataND[0][0]`.

Step 7. Now you will add a microcontroller to your network to control the Coordinator. The following sections use an ARM mbed module or an Arduino Uno module.

USING AN ARM MBED MODULE

Turn off power to the breadboard and leave the End-device module with the resistors and potentiometer attached to it in the breadboard (see Figure 21.1). Remove the Coordinator module from the USB-to-XBee adapter, and set it aside.

If you have a second End-device module in your breadboard, remove it, and set it aside. If you do not have a second adapter socket in your breadboard, insert one and add the LED and resistor shown in Figure 21.5. Make the power (pin 1) and ground (pin 10) connections to the second adapter. If this adapter has any other connections, please remove them now. Connect the ARM mbed module to this adapter and insert the Coordinator module.

If you have set aside a second End-device module, you can place it in another breadboard and add the connections shown earlier in Figure 21.2. You also could place an additional End-device module in the USB-to-XBee adapter. This XBee module would obtain power through the USB cable, but it will not communicate with the PC.

- Turn on power to the Coordinator module and then power the End-device module or modules. After a few seconds you should see the LED on the Coordinator module start to flash, followed by flashes from the LED

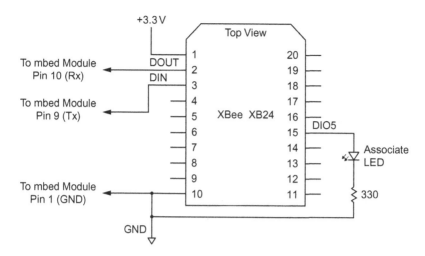

FIGURE 21.5 Circuit for the XBee module used as the Coordinator with an ARM mbed module.

attached to each End-device module. The flashing LEDs indicate the End-device modules have associated with the Coordinator module.

- Program EX21_mbed_ND provides the C-language code you will run on your ARM mbed module.

Program EX21_mbed_ND

Code for an ARM mbed module to control a Coordinator and determine the number of End-device modules you have in a personal-area network (PAN).

```
/*
 * EX21 ARM mbed Module Rev. B
 * Program: EX21_mbed_ND
 * Obtain information about XBee End-device modules in a network,
 * regardless of number of modules. Store information in an array,
 * print hex data.
 * Uses Timer0 and an interrupt
 * Optional LED on pin 13 for debugging and error notification.
 * Jon Titus 09-06-2011 @1010H
 */

#include "mbed.h"

//*************** Declarations & Assignments **************
//Define API "ND" API command-packet array here
int   packetND[] = {0x7E, 0x00, 0x04, 0x08, 0x52, 0x4E, 0x44, 0x13};
int   packetNDlen = 8;

int        XBee_numb;
               //counter for number of module responses
```

```
const int   numb_of_modules = 5;
                //max number of expected modules, change as needed
const int   dataNDlength = 40;
                //max length of dataND array
char        dataND[numb_of_modules][dataNDlength];
                //set up response array for XBee modules

//define additional variables
int   testdata;
int   output_data;
int   packet_start_byte = 0x7E;
DigitalOut    LedPin(p13);
                        //Set pin 13 as indicator LED output
Serial        Sport(p9,p10);
                        //Serial port pins 9 & 10 for XBee
    Serial        pc(USBTX, USBRX);
                        //Set up communications with host PC
unsigned int  timer_count;
unsigned int  bytecount_hi;
unsigned int  bytecount_lo;
unsigned int  bytecount;
unsigned int  counter;
unsigned int  chksum;
unsigned int  chksum_temp;
unsigned int  XBee_total_numb;
unsigned int  Timer0_start_value = 2398000;    //approx. 100 msec.

//************** Set up Timer0 **************
//Set up Timer0 here
void timer0_init(void)
{
    LPC_SC->PCONP |=1<1;   //timer0 power on
    LPC_TIM0->MR0 = Timer0_start_value;
    LPC_TIM0->MCR = 3;     //interrupt and reset control
}                          //3 = interrupt & reset timer0 on match
                           //1 = interrupt only, no reset of timer0
//************** Timer0 Interrupt Svc Routine **************
extern "C" void TIMER0_IRQHandler (void)
                        //compile as a "C" function
{
  if((LPC_TIM0->IR & 0x01) == 0x01)
                        //if interrupt from MR0, then
    {
    LPC_TIM0->IR |= 1 << 0;
                        //clear MR0 interrupt flag
    timer_count++;      //increment timer_counter value
    }
}

//************** Serial-Input Routine **************
//function will read serial port as soon as it has data ready
char SerialInput()
{
    timer_count = 0;        //set outer-loop count
    NVIC_EnableIRQ(TIMER0_IRQn);
                            //enable Timer0 interrupt
    LPC_TIM0->TCR = 1;      //enable Timer0
    while (Sport.readable() == 0)
                //stay in this loop until character received
```

```
    {
    if (timer_count > 80)
                //test outer-loop counter incremented by Timer0 ISR
        {
        //CPU has waited too long for UART response
        NVIC_DisableIRQ(TIMER0_IRQn);
                        //turn off timer interrupt
        return (0);         //abort, return with value of 0
        }
    }
    NVIC_DisableIRQ(TIMER0_IRQn);
                //UART now has data, so turn off timer interrupt
    return (Sport.getc()); //return with UART data
}

//************ Function to send ND API Command Packet ************
//function will send command packet
void SendPacket(int packet2send[], int packetlength)
{
    counter = 0;            //loop counter for bytes in API packet
    while( counter < packetlength )
        {
        Sport.putc(packet2send[counter]);
                        //transmit API-packet byte
        counter++;
        }
}

//************** MAIN PROGRAM ***************

int main(void)
{
    timer0_init();          //initialize Timer0
    LedPin = 1;             //LED at pin 13 off to start
    wait(0.5);              //half-second delay

    SendPacket(packetND, packetNDlen);
                        //Send API packet to XBee Coordinator

    XBee_numb = 0;          //Start with zero XBee End devices

    // This portion of the program gets serial data sent by XBee
    // modules and saves the data in an array of bytes for each module.
    while(1)            //you can only "break" out of this loop
        {
        testdata = SerialInput();
                        //go get UART data, if any
        if (testdata == 0)
                        //SerialInput function aborted--no UART data
            {           //break out of while(1) loop
            break;
            }

        if (testdata != packet_start_byte)
                        //UART has data, so check for start-byte
            {
            while(1)        //First byte is not valid start byte
                {           //Error handling could go here.
                LedPin = 0;
                wait(0.05);
                LedPin = 1;
                wait(0.05);
```

```
         }
      }

else                    //OK, found proper packet-start-byte
      {
      bytecount_hi = SerialInput();
                      //Next two bytes give msg byte count
      bytecount_lo = SerialInput();
      bytecount = (bytecount_hi * 256) + bytecount_lo;
                      //calculate # of bytes in msg
      chksum_temp = 0;
      for (counter = 0; counter < bytecount; counter++)
                      //put all bytes in array until done
        {
        dataND[XBee_numb][counter] = SerialInput();
        chksum_temp = chksum_temp + dataND[XBee_numb][counter];
                      //tally checksum
        }
      chksum = SerialInput();
                      //get checksum--last byte in packet
      if (chksum != (0xFF - (chksum_temp & 0xFF)))
                      //do checksums match?
          {
          while(1)   //no match, flash LED forever
              {      //error handling code could go here instead of
              LedPin = 0; //LED-flash loop
              wait(1);
              LedPin = 1;
              wait(1);
              }
          }
    XBee_numb++;       //increment XBee-numb module count
    }
}                      //get data from next Xbee module
//break out of while(1) loop comes here

//if (XBee_numb > 0) //for the ND command, remove 1 count
//    {               //so final status packet from Coordinator
//    XBee_numb--;    //does not count as a module.
//    }


//Print the number of modules detected and the hex data from
//each discovered
//End-device module.
XBee_total_numb = XBee_numb;
 LedPin = 0;                //turn on external LED
 wait(0.5);                 //half-second delay
 pc.printf("\n\r\n\r");
 pc.printf("Number of modules: ");
                      //print number of discovered modules
 pc.printf("%d  ", XBee_total_numb);
 pc.printf("\n\r");

//loop to print hex data from each End-device XBee module
  //data saved in dataND[][] array
  for (XBee_numb = 0; XBee_numb < XBee_total_numb; XBee_numb++)
  {
```

```
    pc.printf("\n\r");
       for (counter = 0; counter < 40; counter++)
          {
          output_data = dataND[XBee_numb][counter];
          pc.printf("%02X  ",output_data);
                            //print double zeros for zero value
          }
       pc.printf("\n\r");  //go to a new line in text window
    }

    while(1)              //end program in an infinite do-nothing
       {                  //loop. Press reset button to run again
       }
}
//Program ends
//======================================================================
```

- Compile the EX21_mbed_ND program and correct any typing errors. The compiler will download the program into the ARM mbed module as if it exists as an external USB memory stick. On my PC, the ARM mbed module appeared as the F: drive.
  After the LED on the ARM mbed board stops flashing, start your terminal emulator program and set it for 9600 bits/second, 8 data bits, no parity, 1 stop bit, and no flow control. The terminal emulator uses the same USB cable the compiler used to download the code. You can use the Windows Device Manager to determine which virtual serial port the ARM mbed connects to. Change your view so you can see the terminal emulator window.
- Press the reset pushbutton on the ARM mbed module. Remember, the MCU will wait until the timer count ends before it sends the XBee module to your PC. When I ran the code with two end devices, I observed in the terminal emulator window:

```
Number of modules: 3
88 52 4E 44 00 FF FE 00 13 A2 00 40 6A F5 AB 3F 50 52 4F 00 00
00 00...
88 52 4E 44 00 FF FE 00 13 A2 00 40 49 E1 E6 4A 45 4E 44 00 00
00 00...
88 52 4E 44 00 00 00...
```

For clarity I have excluded many 00 values at the end of each data set.
The response from the MCU indicates an "extra" End-device module because the Coordinator sends a short acknowledgement, which the program counts as a module. The software includes four commented-out lines that do not execute:

```
//if (XBee_numb > 0)
//    {
//    XBee_numb--;
//    }
```

To not count the Coordinator's acknowledgement message as an End-device module, "un-comment" these statements so the code decreases the module count by one.

- For the first response, the packet breaks down as follows:

88	= AT Command response
52	= frame-identifier byte (52 for all experiments)
4E44	= hex code ND
00	= status byte (OK)
FFFE	= 16-bit MY value
0013A200406AF5AB	= 64-bit address of responding module
3F	= signal strength
50	= P
52	= R
4F	= O
00	= null, end of NI information

The displayed information came from the `dataND` array. Note the array saved the 64-bit (8-byte) address for each End-device module in locations `dataND[x][7]` through `dataND[x][14]`. The software in Experiment 20 used the SL address bytes, `dataND[x][11]` through `dataND[x][14]` within an IS API command packet to address a specific End-device module. So the `dataND` information gathered here would easily "plug into" the code in Experiment 20 so you could discover modules and obtain analog and digital information from them.

- Turn off power to one of your XBee modules and wait 10 seconds. Now run the program again. Did it detect the proper number of remaining End-device modules? In my lab, I had two End devices, so the program now counted only the single powered XBee End-device module. If you have only one End device and turn off its power you should see: `Number of modules: 0`.

Skip ahead to the Conclusion section at the end of this experiment.

USING AN ARDUINO UNO MODULE

Turn off power to the breadboard and leave the End-device module with the resistors and potentiometer attached to it in the breadboard (see Figure 21.1). Remove the Coordinator module from the USB-to-XBee adapter, and set it aside.

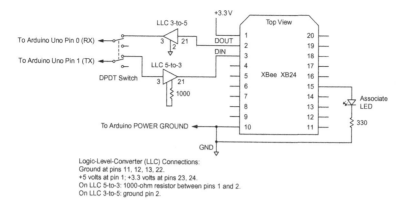

FIGURE 21.6 Circuit for the XBee module used as the Coordinator with an Arduino Uno.

If you have a second End-device module in your breadboard, remove it, and set it aside. If you do not have a second adapter socket in your breadboard, insert one and add the LED and resistor shown in Figure 21.6. Make the power (pin 1) and ground (pin 10) connections, and connect the logic-level-conversion circuits and the Arduino Uno module as shown. If you have other connections to this adapter, please remove them and then insert the Coordinator module.

If you have set aside a second End-device module, you can place it in another breadboard and add the connections shown earlier in Figure 21.2. You also could insert an End-device module in the USB-to-XBee adapter. This XBee module would obtain power through the USB cable, but it will not communicate with the PC.

Pay attention to the notes below and in Figure 21.6 that explain additional connections not shown in the diagram for the sake of clarity:

- On LLC 5-to-3, connect a 1000-ohm resistor (brown-black-red) between pins 1 and 2.
- On LLC 3-to-5, connect pin 2 to ground.
- On both logic-level-converter devices, connect pins 11, 12, 13, and 22 to ground.
- On both logic-level-converter devices, connect pin 1 to +5 volts.
- On both logic-level-converter devices, connect pins 23 and 24 to +3.3 volts.
- Turn on power to the Coordinator module and then power the End-device module or modules. After a few seconds you should see the LED on the Coordinator module start to flash, followed by flashes from the LED attached to each End-device module. The flashing LEDs indicate the End-device modules have associated with the Coordinator module.

Important: As noted in other experiments, the Arduino Uno module shares its UART with an XBee module and the host PC USB connection, so the serial connection between an Arduino Uno module and an XBee module when downloading code can cause an electrical conflict. You MUST manually disconnect (or use a small switch to open) the two serial-port connections between an Arduino Uno module and any external device while the compiler downloads code to the Arduino Uno. I used a small double-pole double-throw (DPDT) toggle switch to disconnect the signals between the logic-level-conversion circuits and the Arduino Uno module during code downloads. Remember to reconnect these wires, or flip the switch, before you run the program.

- Disconnect the wires between the logic-level-converter circuits and the Arduino Uno module RX and TX pins (pin 0 and pin 1). If you don't use a switch, colored wires will help you keep track of the TX and RX connections.

Program EX21_Uno_ND provides the C-language code you will run on your Arduino Uno module.

Program EX21_Uno_ND

Code for an Arduino Uno module to control a Coordinator and determine the number of End-device modules you have in a personal-area network (PAN).

```
/*
 * EX21 Arduino Uno Module   Rev. B
 * Program: EX21_Uno_ND
 * Obtain information about XBee End-device modules in a network,
 * regardless of number of modules. Store information in an array,
 * print hex data.
 * Uses Timer2 and an interrupt.
 * Optional LED on pin 13 for debugging and error notification.
 * Jon Titus 09-05-2011 @1525H
 */

//*************** Declarations & Assignments ***************

// Define API "ND" command-packet array here
byte        packetND[] = {0x7E, 0x00, 0x04, 0x08, 0x52, 0x4E,
                          0x44, 0x13};
byte        packetNDlen = 8;
                //packet ND length in number of bytes

byte        XBee_numb;
                //counter for number of module responses
const byte  numb_of_modules = 5;
                //max number of expected modules, change as needed
const int   dataNDlength = 40;
                //max length of dataND array
byte        dataND[numb_of_modules][dataNDlength];
                //Set up response array for XBee modules

byte        testdata;
                //variable for incoming data
byte        packet_start_byte = 0x7E;
int         LedPin = 13;
                //Set pin 13 as optional indicator LED output

//define additional variables
unsigned int  timer_count;
unsigned int  bytecount_hi;
```

```
unsigned int  bytecount_lo;
unsigned int  bytecount;
unsigned int  counter;
unsigned int  chksum;
unsigned int  chksum_temp;
unsigned int  XBee_total_numb;
unsigned int  Timer2_start_value = 0x02;

//*************** Set up Timer2 ***************
//Set up Timer2
//see Atmel manual for ATmega328P 8-bit MCU family. www.atmel.com
int SetupTimer2()
  {
  TCCR2A = 0;    //Timer2 mode 0, normal operating mode
  TCCR2B = 1<<CS22 | 1<<CS21 | 1<<CS20;
               //clock-select value for divide-by-1024 prescale
  return(0);     //same as TCCR2B = 0x07
  }

//*************** Timer2 Interrupt Svc Routine ***************
//Timer2 overflow interrupt-service routine (ISR)
ISR(TIMER2_OVF_vect)
               //compiler will place ISR at proper location for
  {            //Timer2 overflow interrupt
  TCNT2=Timer2_start_value;
               //reset Timer2 with starting value for counting
               //up to overflow
  timer_count++;
               //increment this variable on each pass through ISR
  }

 //*************** Serial-Input Routine ***************
 //function to read serial port as soon as it has data ready
byte SerialInput()
{
   timer_count = 0;
               //set outer-loop count
   TCNT2 = Timer2_start_value;
               //set timer counter for count up to overflow
   TIMSK2 = 1<<TOIE2;
               //enable timer-overflow interrupt
   while (Serial.available() == 0)
               //stay in this loop until character received
   {           //or until timer_count >255
     if (timer_count > 255)
               //test outer-loop counter incremented by Timer2 ISR
       {       //CPU has waited too long for Coordinator response
       TIMSK2 = 0<<TOIE2;
               //turn off timer-overflow interrupt
       return (0);
       }       //abort SerialInput routine, return with value of 0
   }
   TIMSK2 = 0<<TOIE2;
               //MCU UART now has data, so turn off timer interrupt
   return (Serial.read());
               //return to main program with UART data
}

//************ Setup Routine for Serial, Timer2, LED ************
void setup()
{
```

```
  Serial.begin(9600);
  pinMode(LedPin, OUTPUT);
  digitalWrite(LedPin, HIGH);
            //Turn test LED off to start
  SetupTimer2();
            //execute Timer2 set up
}

//************** Main Loop **************
//MAIN PROGRAM

void loop()
{
  delay(500);    //Half-second delay for switch debounce

  Serial.write (packetND, packetNDlen);
            //Send "ND" API packet to XBee Coordinator

  XBee_numb = 0;
              //Start with zero XBee End devices

  // This portion of the program gets serial data sent by XBee
  // modules and saves the data in an array of bytes for each module.
  //
  while(1)
    {
    testdata = SerialInput();
              //go get UART data, if any
    if (testdata == 0)
              //SerialInput function aborted--no UART data
      {       //BREAK OUT of while(1) loop
      break;
      }

    if (testdata != packet_start_byte)
              //OK, UART has data, so check for start-byte
      {       //if start byte is incorrect
      while(1)    //then run this loop forever
        {       //error handling could go here.
        digitalWrite(LedPin, LOW)
              //flash LED to indicate error
        delay(50);
        digitalWrite(LedPin, HIGH);
        delay (50);
        }
      }
    else          //OK, found proper packet-start-byte
      {
      bytecount_hi = SerialInput();
              //next two bytes give msg byte count
      bytecount_lo = SerialInput();
      bytecount = (bytecount_hi * 256) + bytecount_lo;
              //calculate # of bytes in message
      chksum_temp = 0;
      for (counter = 0; counter < bytecount; counter++)
              //put all bytes in array until done
        {
        dataND[XBee_numb][counter] = SerialInput();
chksum_temp = chksum_temp + dataND[XBee_numb][counter];
              //tally checksum
        }
```

```
        chksum = SerialInput();
                //get checksum--last byte in packet
        if (chksum != (0xFF -(chksum_temp& 0xFF)))
                //do checksums match?
          {
          while(1)
                //no checksum match, flash LED forever
              {   //Error handling could go
              digitalWrite(LedPin, LOW);
                //here instead.
              delay(1000);
              digitalWrite(LedPin, HIGH);
              delay (500);
              }
          }
    XBee_numb++;   //got all data for this XBee module
      }              //increment XBee module count
 }                   //see if you have another module

//BREAK OUT
//break comes to this point
                   //statements commented out--see Experiment text
//if (XBee_numb > 0)
                   //Coordinator includes acknowledgement
//   {             //so decrease module count by one
//   XBee_numb--;  //so Coordinator doesn't count as an
//   }             //End-device module

 XBee_total_numb = XBee_numb;
  digitalWrite(LedPin, LOW);
                   //turn on LED to show data ready
  delay(500);      //half-second delay
  Serial.print("\n\r\n\r");
                   //print module count
  Serial.print("Number of modules: ");
  Serial.print(XBee_total_numb, HEX);
  Serial.print("\n\r");

  //Test routine to print data in each XBee array as hex characters
  for (XBee_numb = 0; XBee_numb < XBee_total_numb; XBee_numb++)
  {
  Serial.print("\n\r");
    for (counter = 0; counter < dataNDlength; counter++)
      {
      if (dataND[XBee_numb][counter] < 0x10)
                   //format single-digit hex values
        {
        Serial.print("0");
        }
      Serial.print(dataND[XBee_numb][counter], HEX);
      Serial.print(" ");
      }
    Serial.print("\n\r");
  }                //go to a new line in text window

  while(1)         //end program in an infinite do-nothing
      {            //loop
      }
}
//Program ends
//===============================================================
```

- Compile the EX21_Uno_ND program and correct any typing errors. Load the code into the Arduino Uno. After you see the message "Done uploading," reattach the wires between the logic-level-conversion (LLC) circuits and the Arduino Uno module RX and TX pins. Do not disconnect the USB cable between the Arduino Uno module and your PC.

  Open the Arduino Uno compiler's Serial Monitor window to see messages produced by the downloaded software. If you have the X-CTU program open, I recommend you close it.

- Press the reset button on the Arduino Uno module to start the program. After the program acquires data from all XBee End-device modules and the `SerialInput` routine times out, the Arduino Uno will send End-device information to the Serial Monitor window. Remember, the MCU will wait until the timer count ends before it sends the XBee module to your PC.

  When I ran the program with two end devices, the Serial Monitor window showed:

```
~...RND.


Number of modules: 3


88 52 4E 44 00 FF FE 00 13 A2 00 40 6A F5 AB 3F 50 52 4F 00 00

00 00...


88 52 4E 44 00 FF FE 00 13 A2 00 40 49 E1 E6 4A 45 4E 44 00 00

00 00...


88 52 4E 44 00 00 00...
```

  For clarity I have excluded many 00 values at the end of each data set. (I used software to format single-digit hex values with a leading zero.)

- Because the Arduino Uno module UART transmits data to both the USB port and to the Coordinator, you see the outgoing ND command along with symbols on the first line in the Serial Monitor window. Next, the program displayed the number of XBee End-device modules in the network followed by the raw information from each one. Remember, you did not preset the number of modules in the code for this experiment.

  The response from the MCU indicates an "extra" End-device module because the Coordinator sends a short acknowledgement, which the program counts as a module. The software includes four commented-out lines that do not execute:

```
//if (XBee_numb > 0)
//    {
//    XBee_numb--;
//    }
```

To not count the Coordinator's acknowledgement message as an End-device module, "un-comment" these statements so the code decreases the module count by one.

- For the first response from my modules, the packet breaks down as follows:

88	= AT Command response
52	= frame-identifier byte (52 for all experiments)
4E44	= hex code ND
00	= status byte (OK)
FFFE	= 16-bit MY value
0013A200406AF5AB	= 64-bit address of responding module
3F	= signal strength
50	= P
52	= R
4F	= O
00	= null, end of NI information

The displayed information came from the dataND array. Note the array saved the 64-bit (8-byte) address for each End-device module in locations dataND[x][7] through dataND[x][14]. The software in Experiment 20 used the SL address bytes, dataND[x][11] through dataND[x][14] within an IS API command packet to address a specific End-device module. So the dataND information gathered here would easily "plug into" the code in Experiment 20 so you could discover modules and then obtain analog and digital information from them.

- Turn off power to one of your XBee modules and wait 10 seconds. Now run the program again. Did it detect the proper number of remaining End-device modules? In my lab, I had two End devices, so the program now counted only the single powered XBee End-device module. If you have only one End device and turn off its power you should see: Number of modules: 0.

CONCLUSION

When you have an unknown number of modules in a PAN, you can use a program such as EX21_Uno_ND or EX21_mbed_ND to issue a command and wait until all modules in range have had an opportunity to reply. The use of a timer and an interrupt lets you set a limit on how long an MCU will wait for responses until it assumes no other XBee modules will associate with the Coordinator.

You can take an alternate approach, though, if you feel uncomfortable with software that includes an interrupt. (Interrupts can prove difficult to debug.)

The ND – Node Discovery command has an associated NT – Node Discover Time number that lets you set the period during which a Coordinator will wait for responses from XBee modules. Periods range from 100 milliseconds (0x01) to 25.2 seconds (0xFC). At the end of this period, the Coordinator sends a short acknowledgement message, `0x88 0x52 0x4E 0x44 0x00 0x00 0x00...` to an MCU. You could write software to examine incoming packets for this series of bytes—or a subset of these bytes—to indicate the "time out" of the node-discovery period. (The software could look for `0x44` in the fourth message byte followed by at least three `0x00` bytes.)

The XBee command set also includes an NO – Node Discover Options command for XBee modules that lets you turn on or off a complete response from the Coordinator that includes the same information as that for an End-device module. Unless you need the Coordinator's address, I recommend you leave the NO option set at 0, its factory default.

Feel free to modify this software and use it as you wish. I hope you will share your code on the http://www.elsevierdirect.com/companion .jsp?ISBN=9780123914040 Web site.

Exploring Cyclic-Sleep Operations

REQUIREMENTS

2 XBee modules
1 XBee adapter board
1 Solderless breadboard
1 USB-to-XBee adapter
1 USB cable—type-A to mini-B
2 LEDs
2 330-ohm resistors, 1/4 watt, 10% resistor (orange-orange-brown)
1 4700-ohm resistor, 1/4 watt, 10% resistor (yellow-violet-red)
1 10-kohm resistor, 1/4 watt, 10% resistor (brown-black-orange)
1 10-kohm potentiometer
Digi X-CTU software running on a Windows PC, with an open USB port

INTRODUCTION

In this experiment you will learn how to use the cyclic-sleep capabilities of XBee modules in a network. These capabilities let you place modules in sleep modes that save power and extend battery life. Keep in mind the cyclic-sleep modes do not let communications occur in real time due to delays as modules wake up from, or remain in, sleep conditions. In remote-controlled robots or medical-patient monitors, for example, you'll probably want real-time operations. In temperature-sensing applications, irrigation controllers, or heating/ventilating equipment, though, changes will occur slowly and you can probably make measurements every 10 to 15 seconds, or with even longer periods between samples. External circuits, though, can cause an immediate wake-up and data transmission. This experiment does not require a microcontroller.

The cyclic-sleep operations offer two ways to communicate between End-device and Coordinator modules in a personal-area network (PAN). In the first case, End-device modules can awaken at fixed intervals, sample I/O pins, and report results to a Coordinator module. In the second case, End-device modules can awaken and poll the Coordinator to ask if it has a command or data for them. If so, the addressed module performs the requested operation. If not, the module goes back to sleep for another set period. You will see how to use both sleep modes in this experiment.

In Experiment 9 you used an external pin to control a hibernate or a doze sleep mode to control how long a module remained in a sleep state. The cyclic-sleep modes explored in this experiment use timers internal to XBee modules to control sleep operations at regular intervals.

Important: After I set up an End-device module in a cyclic-sleep mode for this experiment often it took several attempts to read and write its Modem Configuration information. I suspect the End-device module goes into its sleep mode when powered from a PC through the USB-to-XBee adapter. Just click on the Read or Write button in the X-CTU Modem Configuration window several times to catch the module in its awake state. I don't know a way to overcome this situation.

PERIODIC WAKE-UP WITH I/O REPORT

You might choose to have each End-device module go into a sleep mode and wake-up periodically to report digital or analog information. Assume you want an End-device module to wake up every 15 seconds and transmit information about its analog or digital inputs to the Coordinator module. Your End-device module would use the following settings to establish the sleep mode and its timing:

Sleep Mode (SM). This setting offers two choices for cyclic-sleep operation. You can let an XBee module operate on its own (SM = 4) or you can let it operate on its own but with an external pin wake-up control signal (SM = 5). The external signal will override the internal timing for a *sleeping* End-device module:

0 = Disabled (no sleep mode) default setting
1 = Pin Hibernate (see Experiment 9)
2 = Pin Doze (see Experiment 9)
4 = Cyclic Sleep Remote
5 = Cyclic Sleep Remote with Pin Wakeup

This experiment uses modes 4 and 5. (Digi reserves mode 3 (SM = 3) for a later undefined use.)

Sleep Options (SO). You use this setting to cause the End-device module to sample the active digital and analog inputs and transmit that information upon wake-up.

Cyclic Sleep Period (SP). The SP value determines the sleep period for End-device modules, with a limit of 268 seconds, or 4 minutes and 28 seconds.

Time Before Sleep (ST). An XBee End-device module will go into sleep mode only if it detects no radio communications for this length of time. This value defaults to 5 seconds (0x1388), but you can extend it for as long as 65.5 seconds (0xFFFF). Always use a Time Before Sleep value greater than the Guard Time

(GT), which defaults to one second (0x3E8). The Guard Time establishes the quiet time before and after an AT command. I recommend you not modify the GT setting and always use an ST setting greater than one second.

In this section, you will configure one XBee module as a Coordinator and one as an End device. You may add a second End-device module if you wish, but I recommend you first run the experiment with one module and then go through the experiment again with additional End-device modules, if you have them.

Set Up an End-Device Module

Step 1. Place your End-device XBee module in the USB-to-XBee adapter and use the X-CTU software to Read its Modem Configuration information. (This section uses configuration settings in the file EX22_End_A.pro.)

- Under the heading Networking & Security, ensure you have a value of 0 for CE – Coordinator Enable and a value of 0x1FFE for SC – Scan Channels.
- For the A1 – End Device Association value, select the setting 6 – 0110B, where the B stands for binary. This setting establishes these conditions: associate with a Coordinator on any channel and attempt to associate indefinitely with a PAN Coordinator.
- Set the RN – Random Delay Slots to a value of 2. This setting helps prevent collisions between simultaneous transmissions from several End devices. Even if you have only one End-device module, I recommend you use this setting. (I found errors in transmissions from End-device modules when I used the default value of 0 for RN.)
- Move down to the last item under the Networking & Security heading, NI – Node Identifier, and click on this label. Next, click on the Set button that appears to the right of this label and type in a name for the module. You may use as many as 20 characters, but I recommend you use four or five. (I named my End-device module XMTR.)
- Look under the Sleep Modes (NonBeacon) heading and set the following configurations:
  SM – Sleep Mode   = 4 – CYCLIC SLEEP REMOTE
  ST – Time Before Sleep   = 1388 (default, 5 seconds: 1-msec units)
  SP – Cyclic Sleep Period   = 3E8 (10 seconds: 10-msec units)
  SO – Sleep Options   = 0 (poll and sample)
  Note: The ST parameter uses increments of one millisecond and the SP parameter uses units of 10 milliseconds. Thus, 0x1388 = 5000, which converts to 5000 milliseconds, or 5 seconds. And, 0x3E8 = 1000, which converts to 10,000 milliseconds, or 10 seconds.
- Look under Serial Interfacing heading for the label, AP – API Enable, and click on it. Choose 1 – API ENABLED.
- Write the hexadecimal SH and SL information for your End-device module or modules in Table 22.1. You will need this information later.
  Recheck your Modem Configuration settings.

Table 22.1 End-Device Module Information

	End-Device 1	End-Device 2
SH – Serial Number High		
SL – Serial Number Low		

Note: End-device modules 2 and 3 are optional.

- Move down to the section labeled I/O Settings and change your End-device module settings to match the ones below.

  D8 – DI8 Configuration   0 – DISABLED
  D7 – DIO7 Configuration   1 – CTS FLOW CONTROL
  D6 – DIO6 Configuration   0 – DISABLED
  D5 – DIO5 Configuration   1 – ASSOCIATED INDICATOR
  D4 – DIO4 Configuration   0 – DISABLED
  D3 – DIO3 Configuration   2 – ADC   (Analog-to-digital converter)
  D2 – DIO2 Configuration   3 – DI   (Digital input)
  D1 – DIO1 Configuration   2 – ADC   (Analog-to-digital converter)
  D0 – DIO0 Configuration   3 – DI   (Digital input)
  Recheck your DIO settings.
- Finally, click on Write to save this configuration in the attached End-device module. This step enables the API interface on your module, gives it a name you can recognize, sets it as an End device, establishes sleep operations and times, and enables four I/O pins as inputs.

After you have programmed the End-device module, remove it from the USB-to-XBee adapter and set it aside separate from the XBee module you will use in the next steps.

Set Up a Coordinator Module

Step 2. Place the XBee module you will use as the Coordinator in the USB-to-XBee adapter and use the X-CTU software to Read its Modem Configuration information. (This section uses configuration settings in the file EX22_Coord_A.pro.)

- Ensure you have a value of 0x1FFE for SC – Scan Channels.
- In the line labeled CE – Coordinator Enable, select 1 – COORDINATOR.
- For the A2 – Coordinator Association value, select the setting 6 – 110B, where the B stands for binary. This setting establishes two conditions: It lets the Coordinator find an unused wireless channel and it lets any End device associate with this Coordinator.
- Look under the Sleep Modes (NonBeacon) heading and if necessary, set the following configurations:
  SM – Sleep Mode   = 0 – NO SLEEP (default)
  ST – Time Before Sleep   = 1388 (default, 5 seconds)
  SP – Cyclic Sleep Period = 0 (default)

- Move down to the last item under the Networking & Security heading: NI – Node Identifier and click on this label. Next, click on the Set button that appears to the right of this label and type in a name for the module. You can use as many as 20 characters, but I recommend you use four or five. (I named my Coordinator module RCVR.)
- Look under Serial Interfacing heading for the label, AP – API Enable, and click on it. Choose 1 – API ENABLED.
  Recheck your settings.
- Finally, click on Write to save this configuration in the attached Coordinator module. This step enables the API interface on your module, gives it a name you can recognize, sets the cyclic-sleep times, and sets it as a Coordinator. Note that the Coordinator will NOT go into a sleep state because you set the Sleep Mode to 0.

Leave the Coordinator module in the USB-to-XBee adapter.

Step 3. Place your End-device modules in an XBee adapter in your solderless breadboard and add the components shown in Figure 22.1 to this module. The two digital inputs (DIO0 and DIO2) and two analog inputs (DIO1 and DIO3) provide information this End-device module can transmit to the Coordinator you programmed earlier. This circuit also includes an LED for the module's clear-to-send (CTS) output at pin 12. If the adapter has any other electrical connections, please remove them now.

Step 4. The Coordinator module already has power connected via the USB-to-XBee adapter. Turn on power to your End-device module and note the behavior of the LEDs. What do you observe on the LEDs?

I saw both LEDs turn on and remain on for about four seconds and then turned off. The LEDs remained off for 10 seconds, after which the Associate

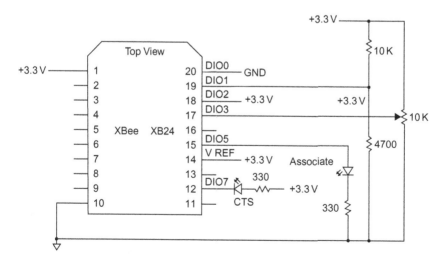

FIGURE 22.1 Add this circuitry to an End-device module so it has two digital and two analog inputs as well as an Associate and a Clear-to-Send (CTS) LED.

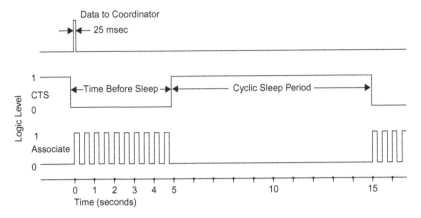

FIGURE 22.2 This diagram shows the relationship between the time an End-device module wakes and sends information to a Coordinator, the Time Before Sleep Period, and the Cyclic Sleep Period.

LED flashed several times and the CTS LED turned on. Both LEDs remained active for five seconds. They repeated this pattern every 15 seconds.

The CTS LED indicates the wake condition for the End-device module. You set a 5-second Time Before Sleep and a 10-second Cyclic Sleep Period. These times sum to 15 seconds. The module requires a 5-second quiet time with no receipt of messages or commands addressed to it before it enters the Cyclic Sleep Period. Thus the Coordinator module receives a transmission about the digital- and analog-input data every 15 seconds, as shown in Figure 22.2.

Important: *An End-device module remains awake during the Time Before Sleep period*, and you can still communicate with it during this time—the associate LED flashes and CTS LED remains on. Each communication that involves this End-device module restarts the Time Before Sleep period. So, you could communicate back and forth with this End-device module for many minutes as long as you never allow more than five seconds between communications. After the final transmission, the End-device module waits for the 5-second Time Before Sleep Period, and absent another communication addressed to it, it goes into the Cyclic Sleep Period.

Open the X-CTU Terminal window. You should see a new packet such as the one that follows every 15 seconds. The packets on my PC looked like this:

```
7E 00 14 82 00 13 A2 00 40 49 E0 EC

3B 00 01 14 05 00 04 01 47 03 62 6D
```

You can change the states of the digital inputs and adjust the potentiometer to see how they affect the information in the received bytes, such as those shown above.

This packet breaks down as follows:

7E	= start byte
0014	= message length (20 bytes)
82	= packet type (Remote Command Response)
00	= frame-identifier byte (52 for all experiments)
0013A2004049E0EC	= 64-bit address of responding module
3B	= signal strength
00	= status byte
01	= number of samples of digital and analog data
14	= first Active-Signal Byte, AD3 and AD1 active
05	= second Active-Signal Byte, DIO2 and DIO0 active
00	= first Digital-Data Byte
04	= second Digital-Data Byte, DIO2 = 1, DIO0 = 0
0247	= hex value from AD1 ADC
0362	= hex value from AD3 ADC
6D	= checksum for this message

The API frame specifications in the Digi International "XBee/XBee-PRO RF Modules" manual do not include information about packet types 0x82 or 0x83. A 0x82 packet type indicates 64-bit addressing for I/O-pin information and a 0x83 indicates 16-bit addressing for I/O-pin information. I found this information in the "API Support" subsection of the "RF Module Operation" chapter.

If you have a second XBee module to use as an End device, follow the programming sequence in Step 1. Then place this module in an adapter in a solderless breadboard and connect pin 1 to +3.3 volts and pin 10 to ground. If you wish you can add the other components shown in Figure 22.1. Turn on power to your second End-device module. You should see data arrive from both modules. Differences in the SL – Serial Number Low bytes lets you tell the difference between data from each End device.

Step 5. Optional: You might want the flexibility to force an End-device module to transmit I/O information based on an external signal, perhaps a signal from an MCU or a switch closure. The Sleep Request input at pin 9 on an End-device module gives you the capability to do this when you program its Sleep Mode (SM) for: 5 – Cyclic Sleep Remote w/ Pin Wakeup.

Program the End-device module with the attached potentiometer for SM = 5. Then make a connection between the End-device module's

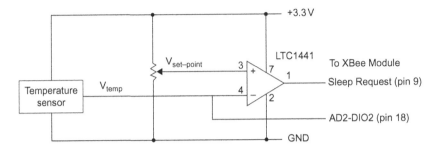

FIGURE 22.3 A basic comparator circuit can trigger the Sleep-Request pin with a logic-1 to a logic-0 transition when the voltage from a sensor exceeds a set-point voltage.

Sleep-Request pin (pin 9) and +3.3 volts on your breadboard. Power the End-device module and let it associate with the Coordinator. After you start to see data in the X-CTU Terminal window, wait until the CTS LED turns off to indicate the module has started a sleep cycle.

Then make a brief connection between ground and the wire that goes to pin 9 on the End-device module. This ground connection changes the logic state at pin 9 to a 0 and triggers the module to wake up and transmit to the Coordinator. (There are better ways to trigger a transmission, but the wire works for a simple experiment.)

Note: The DIO pins on XBee modules have internal pull-up resistors you can enable or disable with the PR setting in the X-CTU Modem Configuration window. A setting of PR = 0x40 would enable the pull-up resistor on pin 9. The XBee/XBee-PRO manual from Digi describes the other pull-up pin settings. The I/O pins default to the pull-up-resistor-on condition: PR – Pull-up Resistor Enable = 0xFF. These resistors will "pull up" an unconnected pin so it assumes a logic-1 state.

For every logic-level *transition* from a logic 1 to a logic 0 at the Sleep-Request pin (pin 9) with the End-device module in its sleep state you should see a response in the Terminal window from the End-device module. But after the XBee module goes into its awake state, logic transitions on the Sleep-Request pin do nothing. *This pin only triggers a sleeping module.* You can't wake up a module that's already awake.

If you have set a long delay between measurements of analog and analog signals, you might want to trigger an extra transmission via the Sleep-Request pin if, say, a temperature exceeds a limit or if someone actuates an emergency switch.

The circuit diagram in Figure 22.3 shows how a temperature-sensor device provides a voltage to one of the ADC inputs on an End-device module, which reports temperature data, say, every 10 minutes. But when the temperature rises past an alarm set-point, you want an immediate report. The LTC1441 comparator continuously compares the sensor voltage to a set-point voltage that represents the upper-limit temperature for the process or equipment monitored by

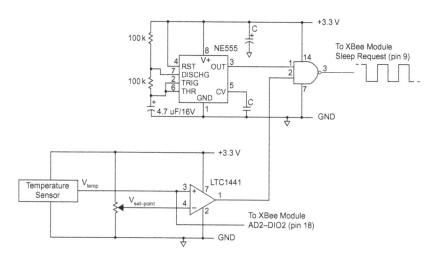

FIGURE 22.4 In this circuit the comparator output switches from a logic 0 to a logic 1 when the voltage from a sensor exceeds a set-point voltage. That logic 1 lets pulses from the NE555-timer circuit trigger the Sleep-Request pin about once a second.

the End-device module. When the sensor voltage exceeds the set-point voltage, the comparator output changes from a logic 1 to a logic 0, which triggers the Sleep-Request pin on the End-device module. The information transmitted to the network Coordinator could immediately alert an attached computer to sound an alarm or take other actions to decrease the temperature.

As shown in Figure 22.4, the comparator might also gate an oscillator built with an NE555-timer IC that would pulse the Sleep-Request pin, say, every second until the voltage from the temperature sensor drops below the set-point voltage. That way packets of I/O information from the End-device module would reach the Coordinator regularly until the temperature decreased below the set point.

Remember that once you wake a module with a logic-level change at the Sleep-Request pin, the module remains awake for the 5-second Time Before Sleep period after it transmits the input-pin information.

If you performed this optional step, please reprogram the End-device module you used so that SM = 4 CYCLIC SLEEP REMOTE. Then proceed to the next section.

PERIODIC WAKE-UP WITH COORDINATOR POLLING

You just learned how to put XBee End-device modules in a cyclic-sleep mode that lets them wake up periodically under their own control and report analog and digital information to a Coordinator module.

You also might need End-device modules that respond with information only when asked, yet still let them take advantage of low-power sleep states. In this cyclic-sleep mode, when an End-device module wakes up periodically it automatically sends a short polling query to the PAN Coordinator

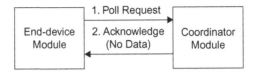

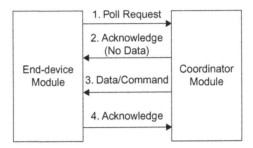

FIGURE 22.5 The sequences in this diagram show how an End-device module will poll a Coordinator when it wakes up and how the Coordinator will respond.

module, as shown in Figure 22.5. If the Coordinator has no command or data for the End-device module the Coordinator module replies with a no-data acknowledgement.

On the other hand, when the Coordinator has a command or data for the awakened End-device module, the Coordinator responds with a data acknowledgement, followed by the command or data in a standard API packet. After the End-device module receives the command, it replies to the Coordinator with an acknowledgement message to signal the command arrived properly. (Digi refers to this as an indirect transmission, but it seems more like a deferred transmission.)

Assume you want an End-device module to wake up every 10 seconds and ask the Coordinator if it has a command for it. The indirect- or deferred-message configurations use the same sleep settings described previously: Sleep Mode (SM), Sleep Options (SO), Cyclic Sleep Period (SP), and Time Before Sleep (ST). Commands directed to a sleeping module must use the 64- or 16-bit address for each End device. Do not try to use broadcast messages with sleeping modules.

In this section, you will configure one XBee module as a Coordinator and one XBee module as End device. Many of the configurations set for the previous Periodic Wake-Up with I/O Report section still apply here.

You may add a second End-device module if you wish, but I recommend you first run the steps that follow with one module and then go through the experiment again with additional End-device modules, if you have them.

Set Up the End-Device Module

Step 6. Place your End-device XBee module in the USB-to-XBee adapter and use the X-CTU software to Read its Modem Configuration information. (This section uses configuration settings in the file EX22_End_B.pro.)

Look under the Sleep Modes (NonBeacon) heading and change the Sleep Options as shown below:

SO – Sleep Options   = 2 (poll, no sample)

Click on Write to save this configuration in the attached End-device module.

After you have programmed the End-device module, remove it from the USB-to-XBee adapter and set it aside and keep it separate from the XBee module you will use in the next steps.

Coordinator Module

Step 7. Place the XBee module you will use as the Coordinator in the USB-to-XBee adapter and use the X-CTU software to Read its Modem Configuration information. (This section uses configuration settings in the file EX22_Coord_B.pro.)

Look under the Sleep Modes (NonBeacon) heading and change the Cyclic Sleep Period as shown below:

SP – Cyclic Sleep Period   = 3E8 (10 seconds)

You might wonder why the Coordinator requires a Cyclic Sleep Period if it will never enter a sleep mode. The Coordinator module should have its Cyclic Sleep Period set to the longest Cyclic Sleep Period of any of the End-device modules in your network. The Coordinator module must know this timing value so it can hold a command or data for an End-device module for *two and a half times* (2.5x) this period, as shown in Figure 22.6. Say you set a 10-second Cyclic Sleep Period for your End-device modules and the Coordinator. The Coordinator will hold for 25 seconds a command or data for an End device. That long period ensures the End-device modules will all wake up at least two times during this 25-second period. If an End-device module does not pick up its message during this 25-second period, the Coordinator purges the waiting command and sends an error message via its UART TX pin. A host microcontroller could interpret this message and take appropriate action.

Recheck your setting. Then click on Write to save this configuration in the attached Coordinator module.

Leave the Coordinator module in the USB-to-XBee adapter.

Step 8. Place your End-device module in the XBee adapter in your solderless breadboard that already has the connections shown earlier in Figure 22.1.

The Coordinator module already has power connected via the USB-to-XBee adapter. Turn on power to your End-device module and note the behavior of the LEDs.

I saw both LEDs turn on and remain on for about four seconds, after which they turned off. After remaining off for 10 seconds the LEDs flashed briefly

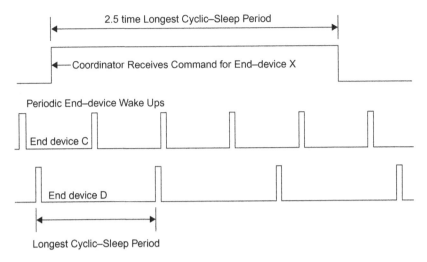

2.5 time Longest Cyclic–Sleep Period

Coordinator Receives Command for End–device X

Periodic End–device Wake Ups

End device C

End device D

Longest Cyclic–Sleep Period

FIGURE 22.6 This diagram shows the times at which the Coordinator receives short polling messages from End-device modules C and D that have different cyclic-sleep periods. Device D has the longest period, so you set the Coordinator's cyclic-sleep period equal to that for Device D. Thus the Coordinator holds data or a command for any End-device module for 2.5 times this period. That extended period allows enough time for all End devices to poll the Coordinator at least twice and receive data or a command addressed to them.

and turned off. They produced a short flash again every 10 seconds, which corresponds to the Cyclic Sleep Period (SP) set for the End-device module. The periodic LED flashes correspond to the time when the module wakes up and polls the Coordinator for any waiting commands. You haven't sent the Coordinator data or a command for the End-device module, so it just goes back to sleep.

Step 9. Open the Terminal window in the X-CTU software, click on Clear Screen, then on Show Hex, and finally on Assemble Packet. Now you will create an API packet for the Force Sample (IS) command to force the End-device module to transmit information about the four input pins you set up earlier. Use the packet framework shown below:

7E 00 0F 17 52 __ __ __ __ __ __ __ __ FF FE 02 49 53   __

Insert the eight address bytes for your End-device module and calculate the checksum. Insert the checksum as the last byte on the right. Go to Table 22.1 for the address information. In the Send Packet window, go to the bottom right corner and select HEX. Type in your hex characters for the API packet.

Step 10. Move your cursor over the Send Data button but do not click it now. Watch the LEDs on the End-device module. A second or two after they turn off, click on Send Data, and watch the X-CTU Terminal window. What happens in the Terminal window?

I saw my transmitted packet and about nine seconds later the window displayed the information transmitted to the Coordinator from my End-device module, as shown in bold type below:

```
7E 00 0F 17 52 00 13 A2 00 40 49 E0

EC FF FE 02 49 53 F1 7E 00 18 97 52

00 13 A2 00 40 49 E0 EC FF FE 49 53

00 01 14 05 00 04 01 47 02 24 E7
```

Based on information in Experiment 20, you can parse the End-device module's information to learn about the state of the digital and analog inputs. Adjust the potentiometer and send the API packet again. Can you see a difference in the data you receive?

Step 11. While the End-device module is in its sleep state (CTS LED off), watch the LEDs as you send the API packet again. What do you observe when the End-device module wakes up?

In my lab I saw the Associate LED flash and the CTS LED remained lit for five seconds. This period corresponds to the Time before Sleep you set at five seconds for the End-device module. So the module remained active for that period. If you transmit the API packet while the CTS LED remains lit, you will see an immediate response to the IS command packet. Now the Coordinator does not have to wait for the module to wake up, so it sends your command immediately.

Step 12. Clear the Terminal window and position the cursor over the Send Data button. Click on the Send Data button and when the CTS LED turns on, click on the Send Data button again and again. You should see many responses in the Terminal Window.

After the End-device module sends the last response, it waits again for five seconds before it goes back into the sleep mode. During its awake period, the End-device module responds immediately. And each communication to or from the End-device module restarts the 5-second timer.

You could use this mode and the IS command when you have devices that only need to report information when commanded by a computer or MCU attached to the Coordinator module's serial port. There's no need to have them wake up and report information if you don't need it. And such periodic wake-ups use precious battery energy.

Step 13. Suppose you send an API command to an End-device module that has a dead battery, has sustained damage, or got moved out of range. What will happen?

Turn off power to your End-device module and clear the X-CTU Terminal window. Now click on Send Data button in the Send Packet window. What happens? You will have to wait for a while to find out.

After 25 seconds, my Terminal window displayed:

```
7E 00 0F 17 52 00 13 A2 00 40 49 E0

EC FF FE 02 49 53 F1 7E 00 0F 97 52

00 13 A2 00 40 49 E0 EC FF FE 49 53

04 6F
```

Why the 25-second delay? The Coordinator module will hold the command packet for the full two and a half times the Cyclic Sleep Period, which you set to 10 seconds. During this period it expects the addressed End-device module to wake up and get its message. But with no End-device module available, the 25-second period times out and the Coordinator responds with an error message. The next-to-the-last byte (0x04 = No Response) in the message above lets you know the Coordinator module had no response from the module with the 64-bit address:

SH = 0013A200 and SL = 4049E0EC

during the 25-second period shown earlier in Figure 22.6. Of course, you can alter the Time Before Sleep and the Cyclic Sleep Period to any times you choose rather than the times used in this experiment.

Important: An XBee Coordinator can hold only two (2) pending messages. As far as I know, you cannot have a microcontroller query a Coordinator to determine how many messages—if any—it has waiting to send to sleeping End devices. Software could increment a count for every message sent to a Coordinator and decrement the counter when the MCU receives a response. That count should give you the number of messages still pending in the Coordinator.

Appendix A
Logic-Level-Conversion Circuits

Most microcontroller (MCU) chips, boards, and modules operate with either a 5-volt or a 3.3-volt power source. The XBee modules, for example, require a 3.3-volt power supply and create logic signals with the following voltages:

3.3-volt logic: Logic 1 = 2.4 to 3.3 volts
                  Logic 0 = 0.0 to 0.4 volts

An MCU module such as the Parallax BASIC Stamp BS2 module and the Arduino Uno board operate with 5-volt logic signals and they operate with signals in the following ranges:

5-volt logic: Logic 1 = 2.4 to 5 volts
                Logic 0 = 0.0 to 0.4 volts

At first glance the logic levels seem somewhat compatible, and you can find schematic diagrams for resistor-only circuits that convert 5V-to-3.3V logic levels. But converting 3V logic signals to 5V signals requires semiconductors. The SparkFun Electronics BOB-08745 board handles four signals, two for 5V-to-3.3V logic signals and two for 3.3V-to-5V logic signals. I have not tried this circuit, though.

While creating the experiments in this book, I used a Texas Instruments SN74LVC4245 "Octal Bus Transceiver" integrated circuit (IC) that shifts the logic level of as many as eight separate signals. The IC lets you choose to have it convert either 3.3V to 5V logic signals or 5V to 3.3V logic levels. Thus you would need two SN74LVC4245 ICs. Because the IC comes in a 24-pin small-outline integrated circuit (SOIC) surface-mount package, it cannot drop into a solderless breadboard with contacts on 0.1-inch centers. A small board such as the SchmartBoard "1.27 mm Pitch SOIC to DIP adapter," part number 204-0004-01, will handle SOIC devices and connect them to pins on 0.1-inch centers. See the Bill of Materials in Appendix F for component information.

I used two such boards, one for 5V-to-3.3V conversions and a second for 3.3V-to-5V conversions. The diagram in Figure A.1 shows the pin numbers and signal names for an SN74LVC4245. You supply +5 volts, +3.3 volts, and ground. The Output Enable (/OE, pin 22) always connects to ground. The Direction (DIR) input connects to 5 volts through a 1000-ohm resistor for

5V-to-3.3V conversions or it connects to ground for 3.3V-to-5V conversions. The "A" side of the SN74LVC4245 IC always operates with 5-volt input or output signals, while the "B" side always operates with 3.3-volt input or output signals. In Figures A.2 and A.3, the large arrow drawn within the 74LVC4245

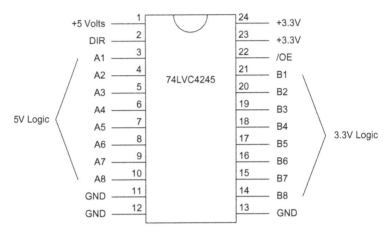

FIGURE A.1 Pin numbers and corresponding signal names for an SN74LVC4245 logic-level-conversion IC in a 24-pin SOIC package.

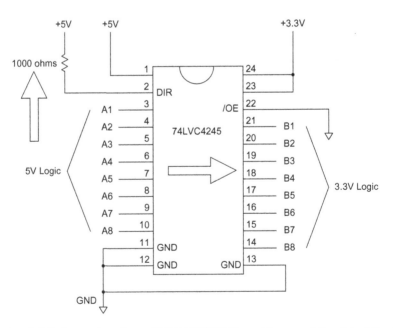

FIGURE A.2 Connections for an SN74LVC4245 IC used to convert 5-volt to 3.3-volt logic levels. Note the 1000-ohm resistor at pin 2 (large upward arrow) and its connection to +5 volts. The right-pointing arrow indicates the direction of signal flow. Always ensure you have a common ground between 5-volt and 3.3-volt power sources.

indicates the signal flow. The up or down arrow points to the direction-control portion of the circuit so you make the proper connection for the type of conversion you need.

In the experiments, you will see the logic-level-converter circuits referred to as "LLC-3-to-5" (three to five) for the 3.3V-to-5V conversions and as "LLC-5-to-3" (five to three) for the 5V-to-3.3V conversions. Again, the "A" side of each LLC *always* operates with 5V signals (in or out) and the B side *always* operates with 3.3V signals (in or out).

For the SN74LVC4245 ICs mounted on adapters I used only 24 header pins and placed the pins and the IC to the far left as shown in Figure A.4. I recommend you make somewhat permanent connections to power and ground, and leave the adapters in a breadboard, as shown in Figure A.5.

The SchmartBoard adapters use a unique way to hold devices in place during soldering and the pads alone provide enough solder. You do need a high-temperature soldering iron with a fine tip. Visit the SchmartBoard Web site for soldering information.

When you solder a board like this to pins on 0.1-inch centers, place the header pins in a solderless breadboard to start, next place the circuit board on the headers, and then solder the pins. This technique properly aligns the pins

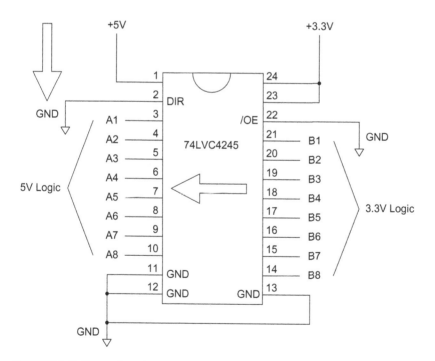

FIGURE A.3 Connections for an SN74LVC4245 IC used to convert 3.3-volt to 5-volt logic levels. Note the ground connection to ground at pin 2 (large downward arrow). The left-pointing arrow indicates the direction of signal flow. Always ensure you have a common ground between 5-volt and 3.3-volt power sources.

FIGURE A.4 A SchmartBoard 28-pin SOIC adapter with an SN74LVC4245 IC and header pins soldered in place.

FIGURE A.5 Two logic-level-conversion circuits in a solderless breadboard. Note the labels that indicate the conversion that occurs in each adapter.

and holds them in place for soldering. Keep the board in the solderless bread-board when you solder the IC pins. It's easier to solder when the breadboard provides stability.

SparkFun Electronics: www.sparkfun.com.

SchmartBoard: www.schmartboard.com.

Appendix B
Hexadecimal Numbers and Checksums

The Digi X-CTU software that configures XBee modules uses hexadecimal values represented by digits 0–9 and letters A–F, which equal the decimal values 0 through 15. This hexadecimal numbering system provides an easy way to represent four-bit binary numbers that comprise only 1s and 0s. Each hex digit represents a 4-bit binary value, from 0000_2 through 1111_2, or from 0x0 to 0xF as shown in Table B.1. The prefix of zero followed by the letter x indicates a value expressed in the hexadecimal numbering system. To represent an 8-bit

Table B.1 Equivalent Decimal, Hexadecimal, and Binary Values

Decimal	Hexadecimal	Binary
0	0	0000
1	1	0001
2	2	0010
3	3	0011
4	4	0100
5	5	0101
6	6	0110
7	7	0111
8	8	1000
9	9	1001
10	A	1010
11	B	1011
12	C	1100
13	D	1101
14	E	1110
15	F	1111

byte, people use two hex digits such as 0x5E, which corresponds to 94_{10}, or 01011110_2. Two-digit hex values range from 0x00 to 0xFF, or in binary from 00000000_2 to 11111111_2, or in decimal from 0 to 255.

Instead of trying to learn how to manipulate hexadecimal numbers, I recommend you buy an inexpensive calculator that can do hex math and convert to and from decimal (base-10) numbers. Texas Instruments and Casio sell calculators with decimal, hexadecimal, and binary modes for under \$US 20. (Practice using a calculator with hex values because you can easily confuse a lower-case B (b) with the numeral 6 on a 7-segment liquid-crystal display. A 6 usually has a top segment while a b does not.)

CHECKSUMS

Many digital-communication protocols sum the bytes in a packet of information to create a checksum that lets a receiving device detect an error. A simple checksum scheme simply adds the bytes in a message and appends the sum at the end of the message. So if you transmit the hexadecimal values:

```
0x09 0xFF 0x4F 0x56 0xA4 0x3B 0x1F
```

you would add the hex values and get a sum of 0x2AB. Because the communication uses bytes, you could send 0x02 followed by 0xAB. Many devices, though, send only the 0xAB byte—the two least-significant hex digits from the sum. The XBee modules do something similar. But instead of sending the entire sum, they take the two least-significant (rightmost) hex digits of the sum—0xAB in the example above—and subtract them from 0xFF to yield a checksum. In the example above, the XBee module would calculate the checksum value as:

```
0xFF - 0xAB = 0x54
```

which would then become the last byte sent, as shown in Table B.2.

An XBee module will automatically reject a command with an incorrect checksum. At times, though, you might need to perform a checksum calculation on incoming data. When that need arises, a microcontroller attached to an XBee module can add the received bytes and use this sum in two ways:

- Take the sum of the received bytes and keep only the two least-significant hex digits. *Subtract* those digits from 0xFF. The result should match the

Table B.2 An Example of a Checksum for a Message Included in a Packet

0x09 0xFF 0x4F 0x56 0xA4 0x3B 0x1F	0x54
Message Bytes	Checksum
Packet	

received checksum value. If it does not, the message contained an error. Again for the message above:

```
0x09 + 0xFF + 0x4F + 0x56 + 0xA4 + 0x3B + 0x1F = 0x2AB
0xFF - 0xAB = 0x54
```

The checksum value you calculate should equal the checksum value appended to the end of the message. If it does not, the message contains an error.

- Take the sum of the received bytes and keep only the two least-significant hex digits. Then *add* them to the received checksum. In the case of the example above:

```
0x09 + 0xFF + 0x4F + 0x56 + 0xA4 + 0x3B + 0x1F = 0x2AB
0xAB + 0x54 = 0xFF
```

This calculation should always yield 0xFF unless the message contains an error. Several software examples in this book use the first method, but you could modify them to use the second method.

Important: The checksum used in XBee command and message packets does not sum all of the bytes in a message or packet. An XBee packet starts with the "start byte" 0x7E followed by two bytes that indicate the number of bytes in the message or command portion of the packet, as shown in Figure B.1. *The checksum involves only the command or message bytes.*

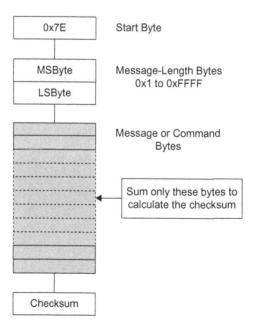

Figure B.1 Checksum calculations for XBee communications do not include the checksum value, the start byte, or the two bytes that indicate the length of a message. Sum only the Message or Command bytes.

A checksum will detect an error in a transmission caused by one bit that changes state, but it cannot tell you which bit or in which byte. Other error-detection and -correction techniques can overcome this shortcoming, but they go beyond the scope of this book. Visit the Wikipedia Web site for an article on "Error detection and correction."

Appendix C
Default Modem-Configuration Settings for XBee (XB24) Modules

Table C.1 Default XBee Modem-Configuration Settings (Firmware version 10E6)

Networking & Security

(C) CH – Channel

(3332) ID – PAN ID

(0) DH – Destination Address High

(0) DL – Destination Address Low

(0) MY – 16-Bit Source Address

(XXXXXXXX) SH – Serial Number High (Depends on module)

(XXXXXXXX) SL – Serial Number Low (Depends on module)

(0) MM – MAC Mode

(0) RR – XBee Retries

(0) RN – Random Delay Slots

(19) NT – Node Discover Time

(0) NO – Node Discovery Options

(0) CE – Coordinator Enable

(1FFE) SC – Scan Channels

(4) SD – Scan Duration

(0) A1 – End Device Association

(0) A2 – Coordinator Association

(0) AI – Association Indication

(0) EE – AES Encryption Enable

KY – AES Encryption Key

( ) NI – Node Identifier

Table C.1 Default XBee Modem-Configuration Settings (Firmware version 1OE6) (continued)

RF Interfacing

(4) PL – Power Level

(2C) CA – CCA Threshold

Sleep Modes

(0) SM – Sleep Mode

(1388) ST – Time Before Sleep

(0) SP – Cyclic Sleep Period

(3E8) DP – Disassociated Cyclic Sleep Period

(0) SO – Sleep Options

Serial Interfacing

(3) BD – Interface Data Rate

(0) NB – Parity

(3) RO – Packetization Timeout

(0) AP – API Enable

(FF) PR – Pull-up Resistor Enable

I/O Settings

(0) D8 – DI8 Configuration

(1) D7 – DIO7 Configuration (CTS Flow Control)

(0) D6 – DIO6 Configuration

(1) D5 – DIO5 Configuration (associated Indicator)

(0) D4 – DIO4 Configuration

(0) D3 – DIO3 Configuration

(0) D2 – DIO2 Configuration

(0) D0 – DIO0 Configuration

(1) IU – I/O Output Enable

(1) IT – Samples before TX

(0) IC – DIO Change Detect

(0) IR – Sample Rate

I/O Line Passing

(FFFFFFFFFFFFFFFF) IA – I/O Input Address

(FF) T0 – D0 Output Timeout

(Continued...)

Table C.1 Default XBee Modem-Configuration settings (Firmware version 10E6) (continued)

(FF) T1 – D1 Output Timeout

(FF) T2 – D2 Output Timeout

(FF) T3 – D3 Output Timeout

(FF) T4 – D4 Output Timeout

(FF) T5 – D5 Output Timeout

(FF) T6 – D6 Output Timeout

(FF) T7 – D7 Output Timeout

(1) P0 – PWM0 Configuration (RSSI)

(0) P1 – PWM1 Configuration

(FF) PT – PWM Output Timeout

(28) RP – RSSI PWM Timer

Diagnostics

(10E6) VR – Firmware Version

(1744) HV – Hardware Version

(0) DB – Received Signal Strength

(0) EC – CCA Failures

(0) EA – ACK Failures

(10000) DD – Device Type Identifier

AT Command Options

(64) CT – AT Command Mode Timeout

(3E8) GT – Guard Times

(2B) CC – Command Sequence Character

Highlighted configuration parameters are read-only.

Appendix D
Electronic and XBee Resources

RESISTORS AND THEIR COLOR CODES

Resistors come in a variety of shapes and sizes, from near-microscopic surface-mount chips to heating elements in electric stoves. The experiments in this book use standard carbon-film 1/4-watt resistors with tolerances of ±5 or ±10 percent. These cylindrical resistors measure about six or seven millimeters in length and two millimeters in diameter and electrical leads extend from each end.

Resistor manufacturers have adopted a standard that uses narrow colored bands to "encode" a resistance value. The colors indicate two digits followed by a multiplier. If you see a resistor marked with yellow-violet-red-gold bands, you can use a color-code chart to determine its value. In this case, yellow = 4, violet = 7, and red = 2 zeros, or 4700 ohms. The gold band indicates a resistance tolerance of ±5%. Find a good color-code chart on the Internet at: www.michaels-electronics-lessons.com/resistor-color-code.html.

You can buy fixed resistors and potentiometers (trimmer resistors) from electronics distributors such as Jameco Electronics, Digi-Key, Allied Electronics, Mouser, and SparkFun Electronics. Resistor assortments include standard values of carbon-film resistors, and those with a ±5 or ±10 percent tolerance will suffice. See the list of suppliers and Web links at the end of this Appendix.

BREADBOARDS

For the experiments I used solderless breadboards that measure about 2.25-by-6.5 inches (5.7-by-16.5 cm). I like the large breadboards because I can fit a lot of circuitry on them and they include power buses. I recommend people insert some 0.1 microfarad (0.1 μF) disc-ceramic capacitors with a rating of at least 20 volts across breadboard power and ground buses. They, and a few 2.2 μF 50V aluminum electrolytic capacitors across the power rails, help attenuate noise spikes. (Electrolytic capacitors have polarity, so connect the + terminal to the + power-supply line.)

WORK AREA

I always use a static-dissipating mat under electronic projects to reduce the possibility of damage caused by electrostatic discharges between me and sensitive components. If you do a lot of electronics work, in the long run a static-dissipating work surface will save you money and frustration. The 3M 8810 ESD Rubber Mat, for example, costs about US$75. You must connect such a mat to a good ground. I also recommend you wear a static-dissipating wrist strap that connects to a snap on the mat. Many distributors sell electrostatic-protection products.

XBee MODULES AND BREADBOARD ADAPTERS

I purchased XBee modules directly from Digi International. Companies such as DigiKey, Parallax, and SparkFun Electronics also sell XBee modules. Review the bill of materials (BOM) in Appendix F to identify the type of module used in the experiment. If you cannot purchase these specific devices, XBee modules for other frequency bands should work equally well and might give you a longer range of wireless communications.

The XBee modules have connectors with pins on 2.0-millimeter centers, but solderless breadboards have receptacles on 0.1-inch centers, so you will need XBee adapters to connect XBee modules to breadboards. I purchased adapters from Parallax and SparkFun Electronics.

The SparkFun adapter, or break-out board (part no. BOB-08276), does not come with 0.1-inch male pins or 2-mm sockets, which you buy separately. This board brings out all 20 pins from an XBee module and leaves enough space so you can make either one or two connections to each pin via a breadboard.

The Parallax adapter (part no. 32403) comes with the needed pins and sockets but it provides 22 pins for your breadboard, which might cause confusion when you count pins on an XBee module and want to make a connection to the Parallax adapter. I made a separate pin-out diagram for the Parallax adapter so I could quickly decide which pins I needed to use. Appendix J provides the XBee module pin-out information. The SparkFun adapter directly follows the XBee module pin numbers.

When you solder pins to an adapter, place the pins in your breadboard and then drop the adapter on them. The breadboard will keep the pins aligned as you solder.

Pay careful attention to module orientation! If you reverse a module when you plug it into an adapter and apply power, you could damage its electronics. The adapters show the orientation of an XBee module. Orient all XBee modules in the same direction. I always put the beveled end of an XBee module to my left.

XBee MODULES AND USB ADAPTERS

The experiments require a USB-to-XBee adapter so you can program configuration parameters in an XBee module and also communicate from a host PC to an XBee module. I purchased two XBee USB Adapter Boards (part no. 32400) from Parallax and used them throughout the experiments. These boards have through-hole solder pads for pins on 0.1-inch centers, so if you wish you can plug them into a solderless breadboard. I didn't use the boards this way, though.

The USB-to-XBee adapter requires a Type-A-to-Mini-A USB cable, through which the XBee module obtains its power from a host PC. Four LEDs on the adapter indicate received signal strength (RSSI), power, Associate, and Sleep. I found the LEDs helpful indicators of XBee operations.

Blue LED   PWM0-Received Signal Strength Indicator (RSSI), XBee pin 6
Red LED   Associate-AD5-DIO5, XBee pin 15
Yellow LED   Power at board
Green LED   ON-Sleep, pin 13

POWER SOURCES

I used two Extech Instruments model 382203 power supplies because I already had them in my lab. You can purchase power supplies with fixed 3.3- or 5-volt outputs or with a variable-voltage output. In most cases a well-regulated "wall-wart" power cube will suffice. You also can buy open-frame power supplies, but I highly recommend you NOT use this type of supply on a lab bench unless you cover the line-power terminals. Otherwise they present a shock hazard. Even some enclosed power-supply modules have exposed connections to line voltage.

Amazon.com lists several suppliers of adjustable power supplies that will power the circuits in this book. Even when an adjustable power supply comes with a meter, I recommend you purchase an inexpensive digital multimeter (DMM) so you can better adjust the power-supply output voltage and also test voltages in your circuits.

As a fall-back, you could use two D-size batteries as a 3-volt power supply and four D cells as a 6-volt power supply. By running the +6 volts through a large diode, such as a 1N5401, you reduce the voltage sufficiently to power 5-volt devices and modules. You can buy plastic battery holders with electrical terminals for a few dollars each.

TOOLS

Soldering requires a soldering iron with a fine tip. I do a lot of soldering and use a Weller WESD51 solder station with variable-temperature control. You not might need a complete station, but you will need a fine tip with a temperature of at least 670 degrees Fahrenheit, or 355 Celsius.

While debugging and testing experiments I used a Saleae "Logic" logic analyzer, which decodes the serial packets transmitted to and from XBee modules and displays them as hex values or ASCII characters. Without the small Logic pod I would not have penetrated some of the mysteries of how the XBee modules operate or communicate. You don't need a logic analyzer to complete the experiments, but keep the inexpensive Saleae unit in mind if you need to troubleshoot digital signals.

I used a Link Instruments MSO-19 combination digital storage oscillo-scope, logic analyzer and pattern generator to capture analog signals and the pulse-width-modulation signals from XBee modules. The MSO-19 and the Logic module connect to a PC via USB cables.

SUPPLIERS

Allied Electronics, www.alliedelec.com

Amazon, www.amazon.com

BusBoard Prototype Systems, www.busboard.us

Digi-Key, www.digikey.com

Digi International, www.digi.com

Jameco Electronics, www.jameco.com

Link Instruments, www.linkinstruments.com

Mouser Electronics, www.mouser.com

Parallax, www.parallax.com

Saleae, www.saleae.com

SchmartBoard, www.schmartboard.com

SparkFun Electronics, www.sparkfun.com

Note: Mention of suppliers and products does not constitute an endorsement. I have no financial interest in any companies or products used or mentioned in this book.

Appendix E
Excel Spreadsheet
Packet-Creator Tool

The Excel spreadsheet Packet Creator 2.xls will help you create command packets of bytes used in several Xbee-module experiments. The spreadsheet converts decimal numbers and ASCII (American Standard Code for Information Interchange) characters into their equivalent hexadecimal values and calculates a checksum. It also helps you determine the byte count in a message and will insert that value in the packet as two bytes.

To create an API command packet that uses an AT command or other information, run Excel and open the Packet Creator 2 spreadsheet (Figure E.1). The sheet has a fixed Start Byte of 0x7E and you will not need to change this value. The spreadsheet provides sample information so you can see how it works.

Do not enter values for message-length values Length MSBy or Length LSBy because the spreadsheet calculates them for you.

Start at Msg Byte Count 1 and insert hexadecimal or decimal values, or ASCII characters for the information you need to include in a command packet. You can mix hex, ASCII, and decimal values as you wish. If you have more than one value in a row, the spreadsheet always uses the left-most value. Suppose you have entered values such those shown in Table E.1.

The first row produces a value of 0x9F, the second 0x5B, and the last 0x57, the hex equivalent of the letter W. To avoid problems, enter only one value in each row.

The Packet column provides the hexadecimal values for your packet. The Decimal column on the right side serves only to compute the checksum. You can ignore or hide this column, but DO NOT delete it.

After you insert all your values for a packet, the spreadsheet computes the checksum. You MUST append the checksum hex value to the end of your packet. The spreadsheet does not do that for you, which prevents you from counting the checksum byte among your message bytes.

Use the numbers in the Msg Byte Count column to determine the length of your message and insert that value in the Message Bytes box at the bottom of the spreadsheet. The proper hex values appear automatically in the Length MSBy and Length LSBy cells at the start of your packet.

Feel free to change or improve this spreadsheet as you wish. I provide it under the license arrangements described in the Introduction to this book. If you want to share your changes, let me know and I'll post them for others.

	Msg Byte Count	Hexadecimal	ASCII	Decimal		Packet	Decimal
Start Byte		7E				7E	126
Length MSBy				0		00	0
Length MSBy				16		10	16
API Identifier	1	17				17	23
	2	52				52	82
	3	00				00	0
	4	00				00	0
	5	00				00	0

Figure E.1 This portion of the Packet Creator 2 spreadsheet shows the columns in which you may place hexadecimal or decimal values, or ASCII characters.

Table E.1 Examples of Entries in the Excel PacketCreator Spreadsheet

Hexadecimal	ASCII	Decimal
9F	A	
5B		21
	W	125

TROUBLESHOOTING

If you run into results that display #NAME? in a cell, install and load the Analysis ToolPak add-in for Excel.

Appendix F
XBee Experiments Bill of Materials

Line	Minimum Qty	Recommended Qty	Description	Supplier	Part No.
1	2	3	XBee XB24 module (XB24-ACI-001)	Digi International Parallax SparkFun Electronics	X24-ACI-001 32406 WRL-08664
2	1	1	USB-to-XBee adapter	Parallax SparkFun Electronics	32400 WRL-08687
3	2	3	XBee adapter	Parallax SparkFun Electronics	32403 BOB-08276
4	1	1	USB cable, type-A to mini-B	Local supplier	
5	1	2	Power supply, 3.3 volts	DigiKey	285-1887-ND
6	1	2	Battery holder, two D cells	DigiKey	176K-ND
7	1	1	Power supply, 5 volts	DigiKey	285-1890-ND
8	1	1	Battery holder, four D cells	DigiKey	BH24DL-ND
9	2	3	Solderless breadboard	DigiKey Jameco Electronics SparkFun Electronics BusBoard Prototype Systems	438-1045-ND 2125026 PRT-00112 BB830
10	100-ft	100-ft	Wire, 22- or 24-gauge, solid conductor	Jameco Electronics	
11	5	10	LED, any color	DigiKey Jameco Electronics Parallax SparkFun Electronics	
12	1	1	Small screwdriver, flat blade	Local hardware store	
13	1	1	Pushbutton, normally open	Jameco Electronics	26623

(Continued ...)

Line	Minimum Qty	Recommended Qty	Description	Supplier	Part No.
14	1	1	Switch, double-pole, double throw (DPDT)	Jameco Electronics	21977
15	10	10	220-ohm, 1/4-watt, 10% resistor	Jameco Electronics	
16	10	10	330-ohm, 1/4-watt, 10% resistor	Jameco Electronics	
17	10	10	1000-ohm, 1/4-watt, 10% resistor	Jameco Electronics	
18	10	10	4700-ohm, 1/4-watt, 10% resistor	Jameco Electronics	
19	10	10	10-kohm, 1/4-watt, 10% resistor	Jameco Electronics	
20	1	1	10-kohm variable resistor (trimmer)	Jameco Electronics	43001
22	1	1	Arduino Uno microcontroller module	SparkFun Electronics Jameco Electronics	DEV-11021 2121105
23	1	1	ARM mbed microcontroller module	DigiKey SparkFun Electronics	568-4916-ND DEV-09564
24	2	2	SOIC adapter boards	SchmartBoard	204-0004-01
25	2	2	SN74LVC4245 integrated circuit	DigiKey	296-14911-1-ND
26	1	10	39-kohm, 1/4-watt, 10% resistor	Jameco Electronics	
27	1	1	Photocell, CdS, 23 to 33 kohms	DigiKey	PDV-P9003-1-ND

Notes by part line number:

All	Part numbers and information are current as of February 2012. From time to time, suppliers drop products or do not have parts in stock. You can find components and supplies at other companies. I note these specific companies because I have purchased products from them and found them reliable and fair.
3	The SparkFun adapter requires purchase of separate receptacles, PRT-08272, and pins, PRT-00116.
5, 6	You can use two D-size 1.5-volt dry cells in series to provide power for 3.3-volt devices if you don't want a plug in power supply.
7, 8	You can use four D-size 1.5-volt dry cells in series to provide a 6-volt power supply. Put a 1N5401 diode in series with the 6 volts and you will have about 5.25 volts, which will not damage 5-volt devices. You only need a 5-volt supply if you plan to use an Arduino Uno microcontroller board.
5, 6, 7, 8	SparkFun sells a small power-regulator that delivers either 5 or 3.3 volts from a wall-wart-type power unit. If you need a maximum of one ampere, these $10 kits deserve a look. SparkFun part no. PRT-00114.
10	Jameco offers the best price on 100-foot spools of 22-gauge hookup wire. I recommend you use two or three colors.
11	You can buy small LEDs from many suppliers. I recommend the 5- or 3-mm-diameter LEDs.
15-19, 25	Rather than buy small quantities of individual resistor values, look for an assortment, such as the Jameco part no. 81832. The experiments specify 10-percent-tolerance resistors, but 5-percent-tolerance resistors are fine.
23	Look under the "SMT to DIP Adapters" heading. You need only one of these microcontroller boards to perform the experiments. I preferred the ARM mbed because it does not need logic-level-conversion circuits and has several serial ports, which simplified experiments.
25, 26	Optional for Experiment No. 6.

Appendix G
American Standard
Code for Information
Interchange (ASCII)

Table G.1 Decimal and Hexadecimal Values for ASCII Characters

Decimal	Hex	Character	Decimal	Hex	Character	Decimal	Hex	Character	Decimal	Hex	Character
0	0	NUL	32	20	space	64	40	@	96	60	`
1	1	SOH	33	21	!	65	41	A	97	61	a
2	2	STX	34	22	"	66	42	B	98	62	b
3	3	ETX	35	23	#	67	43	C	99	63	c
4	4	EOT	36	24	$	68	44	D	100	64	d
5	5	ENQ	37	25	%	69	45	E	101	65	e
6	6	ACK	38	26	&	70	46	F	102	66	f
7	7	BEL	39	27	'	71	47	G	103	67	g
8	8	BS	40	28	(	72	48	H	104	68	h
9	9	TAB	41	29	)	73	49	I	105	69	i
10	A	LF	42	2A	*	74	4A	J	106	6A	j
11	B	VT	43	2B	+	75	4B	K	107	6B	k
12	C	FF	44	2C	,	76	4C	L	108	6C	l
13	D	CR	45	2D	-	77	4D	M	109	6D	m
14	E	SO	46	2E	.	78	4E	N	110	6E	n
15	F	SI	47	2F	/	79	4F	O	111	6F	o
16	10	DLE	48	30	0	80	50	P	112	70	p

Dec	Hex	Char	Dec	Hex	Char	Dec	Hex	Char	Dec	Hex	Char
17	11	DC1	49	31	1	81	51	Q	113	71	q
18	12	DC2	50	32	2	82	52	R	114	72	r
19	13	DC3	51	33	3	83	53	S	115	73	s
20	14	DC4	52	34	4	84	54	T	116	74	t
21	15	NAK	53	35	5	85	55	U	117	75	u
22	16	SYN	54	36	6	86	56	V	118	76	v
23	17	ETB	55	37	7	87	57	W	119	77	w
24	18	CAN	56	38	8	88	58	X	120	78	x
25	19	EM	57	39	9	89	59	Y	121	79	y
26	1A	SUB	58	3A	:	90	5A	Z	122	7A	z
27	1B	ESC	59	3B	;	91	5B	[	123	7B	{
28	1C	FS	60	3C	<	92	5C	\	124	7C	|
29	1D	GS	61	3D	=	93	5D	]	125	7D	}
30	1E	RS	62	3E	>	94	5E	^	126	7E	~
31	1F	US	63	3F	?	95	5F	_	127	7F	DEL

Table G.2 Non-Printing ASCII Information and its Corresponding Action

Symbol	Action
ACK	Acknowledge
BEL	Bell (ring a bell at the receiver)
BS	Backspace
CAN	Cancel
CR	Carriage return
DC1	Device Control 1
DC2	Device Control 2
DC3	Device Control 3
DC4	Device Control 4
DEL	Delete
DLE	Data link escape
EM	End of message
ENQ	Enquire
EOT	End of transmission
ESC	Escape
ETB	End of text block
ETX	End of text
FF	Form feed
FS	File separator
GS	Group separator
LF	Line feed
NAK	Negative acknowledge
NUL	Null
RS	Record separator
SI	Shift in
SO	Shift out
SOH	Start of header
STX	Start of text
SUB	Substitute
SYN	Synchronous idle
TAB	Tab
US	Unit separator
VT	Vertical tab

Appendix H
Troubleshooting

No matter how carefully you construct an electronic circuit, mount components, and create software, you will run into problems, probably more often than you like. The following suggestions will help you overcome problems that might arise as you perform the XBee-module experiments in this book. I base these suggestions and thoughts on experience.

Many electrical problems stem from poor power supplies. Always ensure that you have an adjustable power supply properly set for the voltage or voltages you need. Even though a supply might have terminals marked for, say +5 volts, for a fixed output, have a digital multimeter handy so you can measure voltages at power-supply terminals. Power supplies can have problems, too.

I like to have an LED on breadboards to indicate the presence of power. A quick look lets me know I have power available. Remember an LED needs a current-limiting resistor.

Use wire with sufficient current-carrying capability to not cause a voltage drop between a power supply and a circuit. I have seen some power leads that looked robust, but their thick insulation held fairly thin conductors.

Even if you have power at a power supply and breadboard, it's easy to forget to provide power and ground to a circuit. Always double check power connections to circuits, XBee adapters, and components.

Power supplies that connect to a circuit require a common ground that gives all of the components a common reference of zero volts. Many problems arise because people forget to provide such a ground. I recommend you use one point in a circuit as a common ground point rather than run ground wires here and there. The common-ground point also makes it easy to connect a voltmeter, logic analyzer, or oscilloscope ground lead and make measurements referenced to the same point.

In the bill of materials I recommend using solid-conductor hookup wire of several colors. Engineers commonly use red for power and black for ground. In my lab, I have pre-stripped blue, yellow, and orange wires of several lengths so I can assign one color to one type of signal. Keep your connections short.

On solderless breadboards I usually have $0.1\,\mu F$ disc-ceramic capacitors along with a few 2.2 or $4.7\,\mu F$ electrolytic capacitors that can help absorb power glitches common in circuits. I place these capacitors at intervals between ground and power on a breadboard. Some breadboards have "split" power rails that go halfway across the top or bottom. Remember to use a short jumper to bridge these gaps. Otherwise you'll wonder why one side of the breadboard has power and the other doesn't.

When you plug components into solderless breadboards, ensure pins and wires go into the receptacles. I have seen dual inline-package integrated circuits pressed into a breadboard but with a lead bent up under the IC body. The IC appears properly inserted, but one lead never connects with the breadboard conductor. The same sort of thing can happen with discrete components. They might appear inserted in a breadboard receptacle, but a lead might not contact a conductor.

Integrated circuits, transistors, diodes, XBee modules, and other semiconductor devices can suffer irreversible damage if you reverse their power connections. Power "reversal" comes from human error—either improper connection of power from a power supply, incorrect power connections in a breadboard, or improper insertion of a device in a breadboard or socket. Always insert semiconductor devices with power off and carefully check power connections and device orientation. I once saw a series of 5-volt integrated circuits go up in smoke because someone improperly connected power and ground to signal inputs. In the early 1970s, I burned out a microprocessor chip by making wrong power connections. That was an expensive lesson—the chip cost $360!

Because XBee devices use wireless communications, ensure you have a clear area around their antennas. Placing metal objects near an antenna will affect its radio-frequency radiation pattern and might even shield an antenna completely. I have heard several stories about equipment designers who enclosed a GPS antenna within a metal chassis. Obviously, the GPS receiver did not pick up any satellite signals.

I have had a few cases of "balky" XBee modules. Apparently when we configure XBee modules for sleep modes, they stay in those modes when placed in a USB-to-XBee adapter. Thus it can take several attempts to communicate with a module via the Digi International X-CTU software. Don't give up and think you have a "dead" module. Eventually you will connect with the XBee module and can restore the default XBee settings.

If you doubt the configuration information in an XBee module, you can almost always connect it to a PC that runs the X-CTU software, read the configuration setup, compare it with what you expect, and make changes. You also can restore the factory-default settings.

The X-CTU software lets people update the XBee-module firmware when they save configuration information in a module. I recommend against doing so because a "partial" upgrade—one interrupted part way through—can render a module inoperative. Remember, internal software controls an XBee module and you don't want to mess it up.

If you truly cannot communicate with an XBee module via the X-CTU software, try a new USB cable. We tend to take cables for granted and they can get damaged when run underneath computer cases, run over by chair wheels, stepped on, and so on. Some imported USB-cable manufacturers use cheap, thin wires and pay little attention to quality. For lab use, brand-name USB cables usually do not fail.

One problem I ran into several times involved changing a packet to send in the X-CTU program and forgetting to change the number-of-bytes information at the start of the packet. If you modify a packet, change the number of bytes accordingly. Also, do not add the start byte or the following two bytes that provide the number of bytes in the message portion of a packet when you calculate the checksum. The Packet Creator 2 Excel spreadsheet can help bypass this problem.

Several experiments use microcontroller boards or modules. Always ensure you connect them to a common ground. The small receptacles on an Arduino Uno and the tiny pins on an ARM mbed make it a challenge to connect wires. I like the small grabber-type clip leads for those types of connections. Before you connect anything to serial inputs and outputs, make a diagram that shows how the serial signals will go out and in. It's easier than you think to confuse a serial input with a serial output and mix up what signal goes where. Serial communications can cause quite a few problems.

The default XBee-module configuration has UARTs set for 9600 bits/second, 8 data bits, no parity bit, and 1 stop bit noted in engineering jargon as "8N1." Always ensure you have the same 9600 bits/second, 8N1 settings at both ends of a serial-communication link. Some terminal-emulator programs and some serial-port software lets users set "flow control" so hardware or software can indicate when it's ready for communications. These experiments assume you have set any flow-control settings to "NONE." An incorrect flow-control setting caused me an hour of grief one evening.

The XBee modules can communicate at a higher bit rate, but I recommend you stay with the 9600 bits/second rate. I did not use a higher rate in the experiments.

If you do not program in the C language, I recommend two books:

- "C in a Nutshell," by Peter Prinz and Tony Crawford," O'Reilly Media, ISBN: 978-0596006976, and,
- "Practical C Programming," by Steve Oualline, O'Reilly Media, ISBN: 978-1565923065.

I used both books as references when working with the Arduino Uno and ARM mbed modules. My work doesn't involve full-time programming, so I needed some refreshing in C syntax and structure.

Many software errors result from poor logic; that is, a poor understanding of what we want a computer chip to do. The chips do exactly what we program them to do, but our program errors lead them astray. I often uncovered the following problems in the code I wrote:

- Typing errors. As a two-finger typist I do fairly well, but sometimes I hit two keys or hit the wrong key.
- Syntax errors. I would forget the occasional brace, {or}, in a program that would cause errors. Leaving a semicolon off the end of a C statement can cause compilation errors and a compiler may not detect the error until well

past the error. Don't rely on a compiler to precisely identify the location of a problem.

- Undeclared variables and constants. I got better as I went along, but sometimes I thought I had declared a variable only to find I hadn't.
- Off-by-one errors. In a conditional statement it's easy to cause off-by-one errors. Do I want this loop to end when counter equals 15 or 14? Is the last element in an array data [76] or data [77]?

When writing MCU software I find it handy to have a couple of LEDs connected to output pins and a pushbutton connected to an input pin. I can use the LEDs to let me know I reached a certain point in code and I often used short loops to flash an LED at different rates, depending where the program got to. In several experiments I left an LED and "switch" in the code so you know when software reached a given point, or when it detected an error. That type of debugging aid proved invaluable several times.

If you run out of troubleshooting or testing ideas, take a break and come back to your lab bench with a fresh outlook. Then look at your hardware and software and ask yourself, "What does the result tell me?" In one case I could not get a remote XBee module to respond to any command I sent it. I knew the XBee module didn't have a problem because I could put it in a USB-to-XBee adapter and communicate with it and get it to respond to commands from the X-CTU program. So I put the XBee module back in its remote breadboard, poured a glass of iced tea and read a magazine for 15 minutes. When I returned to my lab bench and reviewed the problem—absolutely no response from the remote module—it dawned on me I sent AT commands to the local XBee module connected to my PC instead of sending remote AT commands. When I corrected the command packets, the XBee module across the room responded properly.

Appendix I
Blank Tables

The blank tables in this appendix let you enter information from your experiments and interpret the results. Feel free to make copies of the tables so you have them in a form convenient to your work.

Table I.1 Active-Signal Byte Table

| | First Active-Signal Byte | | | | | | | |
	First Hex Character				Second Hex Character			
Bit Position	B7	B6	B5	B4	B3	B2	B1	B0
Bit Function	X	A5	A4	A3	A2	A1	A0	D8
Data								

| | Second Active-Signal Byte | | | | | | | |
	First Hex Character				Second Hex Character			
Bit Position	B7	B6	B5	B4	B3	B2	B1	B0
Bit Function	D7	D6	D5	D4	D3	D2	D1	D0
Data								

Table I.2 Digital-Data Byte Table

First Digital-Data Byte							
First Hex Character				Second Hex Character			
Bit Position	B7	B6	B5	B4	B3	B2	B1	B0
Bit Function	X	X	X	X	X	X	X	D8
Data								

Second Digital-Data Byte							
First Hex Character				Second Hex Character			
Bit Position	B7	B6	B5	B4	B3	B2	B1	B0
Bit Function	D7	D6	D5	D4	D3	D2	D1	D0
Data								

Table I.3 Analog-Data Byte Table

	First Analog-Data Byte							
	First Hex Character				Second Hex Character			
Bit Position	B7	B6	B5	B4	B3	B2	B1	B0
Bit Function	X	X	X	X	X	X	A9	A8
Data								

	Second Analog-Data Byte							
	First Hex Character				Second Hex Character			
Bit Position	B7	B6	B5	B4	B3	B2	B1	B0
Bit Function	A7	A6	A5	A4	A3	A2	A1	A0
Data								

Appendix J
XBee Connection
Information

The diagrams in this appendix provide a pin-out drawing for an XB24 XBee module (Figure J.1), and a drawing that labels all XBee signals (Figure J.2). Feel free to copy the first and use it to sketch circuit diagrams. You also can copy the second diagram and keep it near your breadboards to make it easier to find the proper pins and signals.

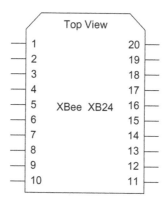

Figure J.1 A pin-out diagram for use in XBee schematic diagrams.

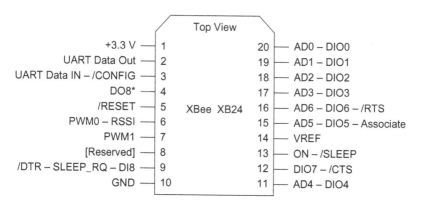

* DO8 not supported at this time.

Figure J.2 This diagram labels the pins on an XBee module to make them easy to identify.

293

The software files you can download from www.elsevierdirect.com/companions.jsp?ISBN=9780123914040 include an XBee-module component for the free schematic-capture software available from ExpressPCB (www.expresspcb.com). This custom component simplifies drawing circuits that require an XBee module.

Glossary

ADC see Analog to Digital Converter

American Standard Code for Information Interchange (ASCII) a code used to communicate 128 unique letters, numbers, punctuation, symbols, and special commands. Although the original ASCII code used only seven bits, the extended ASCII code uses eight bits and includes non-Roman letters and currency symbols.

Ampere a unit of measure for the quantity of current that flows in a circuit and a basic International System of Units (SI) measurement. Named after the French physicist and mathematician André Marie Ampère (1775–1836).

Analog in electronics, a type of signal that can vary without discrete steps.

Analog to digital converter (ADC) an electronic circuit that converts an analog signal into a value represented by n discrete steps. The number of steps and the ADC reference voltage determine the step size and input-measurement range. A 10-bit ADC, for example, has a resolution of 2^{10} (1024_{10}) values, for a theoretical accuracy of 1 part in 1024, or 0.097 percent. Resolution, though, does not directly represent accuracy. In practice, an ADC with a 10-bit resolution offers eight or nine bits of accuracy due to electrical noise and other electronic effects. An ADC requires a stable internal or external reference voltage upon which it bases the conversions and which establishes the range for the converter.

Anode The positive terminal of a power source or an electronic component that has a polarity; that is, a positive and a negative terminal. An LED, for example, has an anode that connects to a positive voltage and a cathode that connects to a negative voltage, or ground (zero volts). See also Cathode.

API see Application Programming Interface

Application programming interface (API) a set of high-level programming functions that automatically perform complicated operations without involving a programmer in their details. Manufacturers of microcontrollers and software suppliers offer APIs specific to their products. An API function such as `read_port_C` used in a program might handle all the steps needed to read values from Port C on a microcontroller. A programmer can use this API function without having to know what connects to Port C, how to control Port C, or what other parts of a program also use Port C.

ASCII see American Standard Code for Information Interchange

Asynchronous the lack of coordinated timing between events, or the lack of a common clocking signal for electronic events.

AT command A set of lettered commands established in the early 1980's to control dial-up modems. A modem would respond to commands that started with the letters AT, which stands for "attention." Over the years, communication-device manufacturers created their own sets of AT commands, such as ATFR for a forced reset, or ATID for "identify yourself." These types of commands simplify control of communication devices such as XBee modules.

Baud Now a synonym with bits per second (bits/sec.), a measure of the speed of asynchronous communications between electronic equipment. Named after Jean Baudot, who invented the Baudot code, an early 5-bit representation of, letters, numbers, and punctuation.

Binary a numbering system that uses only one of two values, 0 or 1, in each numeral position. Binary numbers use 2 as their base, which means positions indicate 16's, 8's, 4's, 2's, and 1s. Thus the binary number $0101_2 = 4 + 1$ or 5_{10}. Each position represents 2^n, where n increases by 1 as bit positions move to the left, as in $2^5, 2^4, 2^3, 2^2, 2^1, 2^0$.

Bit a single-digit binary value of either 1 or 0.

Breadboard originally a piece of wood on which hobbyists, students, and engineers constructed electrical circuits by attaching them to the board, perhaps even a bread breadboard. Present breadboards use plastic strips with rows and columns of metal receptacles, a printed-circuit board with etched pads that allow for soldered connections, or perforated boards in which people insert components, sockets, and wires.

Byte a group of eight continuous bits, such as 10110001_2. A byte has a most-significant bit (MSB) on the left because its bit position has the largest value of all eight bits, either 0 or 128 (2^7). Conversely, the least-significant bit (LSB) on the right contributes the smallest value; either 0 or 1 (2^0). The byte shown here has a value of 177_{10}. See also Binary and Bit.

Capacitor an element that stores charge.

Cathode The negative terminal of a power source or an electronic component that has a polarity; that is, a positive and a negative terminal. An LED, for example, has a cathode that connects to a negative voltage, or ground (zero volts), and an anode that connects to a positive voltage. See also Anode.

Checksum a value computed from information contained in a defined packet and sent with that packet to help a receiver detect an error that occurred during transmission. The receiving device can calculate a checksum in the same manner and compare it to the received checksum. A difference in the checksum usually indicates an error. Modern techniques better detect errors and help correct them, but for short communications, a checksum usually suffices.

COM port a communication port on a personal computer, usually a serial port labeled "10101" near a connector on the back or side of a computer. Many new computers do not provide a COM port, but instead employ a USB port as a virtual serial port that lets application software "think" it communicates with a serial port. The receiving device includes a USB-converter integrated circuit the receiver "sees" as a serial port that connects to a PC. COM ports have a numeric designation, such as COM0 for a built-in serial port, and, say, COM19 for a USB-to-XBee adapter.

Comparator an electronic component that compares two voltages. A comparator comes in handy when you only need to know if one voltage rose above another or dropped below another. A comparator has one output and a plus and a minus input. When input voltage V− exceeds the voltage at the V+ pin, the comparator output changes from a logic 1 to a logic 0.

Digital in electronics, a signal that can exist in discrete voltage steps or discrete values.

Firmware a computer program and data held semipermanently in a microcontroller or similar device. The firmware governs how the device operates. Some applications require an occasional or periodic firmware update, say, a postal meter that needs new postal rates, or an automated toll-collection booth that needs new information about tolls.

Ground an electrical or electronic reference point that represents zero volts in a circuit. Distinct from a ground used in household line-power circuits that include an earth ground.

Hexadecimal a numbering system that represents a digit with a single symbol for the 16 values 0 through 15, with numerals 0 through 9 and letters A through F. The numerals 0 through 9 equal their assigned value, and letters A through F represent values 11 through 15. Hexadecimal numbers simplify the notation of byte values, which can split into two portions, for example, 11000111_2, or 1100 0111. The maximum value for four bits comes to 15, so the binary number 1100 0111 (artificially divided for clarity) becomes C7. Hexadecimal numbers in computer programs have a zero-x prefix, thus 0xC7. Also referred to as hex.

I/O shorthand notation for input/output. Often used in "I/O port" to designate an electronic connection or connections used by a microcontroller or similar device to communicate with electronic devices.

Interrupt an immediate-attention condition caused by hardware or software that stops normal program flow as a processor branches off to handle the cause of the interrupt. An external device could cause an interrupt because it has data the processor must immediately respond to, or software could cause an interrupt if an error occurs, perhaps from a divide-by-zero operation. See also Interrupt-Service Routine.

Interrupt-service routine (ISR) software separate from the normal program flow and written specifically to handle an interrupt for a specific device. An internal timer would have its own ISR, separate from an ISR written to handle an interrupt from an analog-to-digital converter. See also Interrupt.

ISR see Interrupt-Service Routine

Jumper a piece of wire, usually insulated, that connects two points in a circuit. Often a temporary connection.

LED see Light-emitting diode

Light-emitting diode a semiconductor device that emits light with an intensity that depends on the current flow through it. LEDs usually have a narrow wavelength emission, thus you can find, red, green, orange, yellow, and blue LEDs. LEDs also can emit light at infrared or ultraviolet wavelengths.

Logic-level converter a circuit designed to convert the logic-level voltages for one family of devices to the levels appropriate for another family. The 3.3-volt logic in an XBee module cannot directly connect to a microcontroller that uses 5-volt logic. Special integrated circuits perform the conversions to and from 3.3- and 5-volt logic families.

Low-pass filter an electrical or electronic circuit that allows only low-frequency signals to pass. Filter-design software and formulas simplify the design of such filters. You also can have high-pass filters. A band-pass filter allows a specific continuous range of frequencies to pass. A notch filter acts to block signals at a specific frequency.

MCU see Microcontroller

Microcontroller an integrated circuit that provides a central processor that performs math and logic operations and can move information from place to place. The microcontroller includes memory for storage of program code and to store information. Peripheral devices within the IC include timers, input-output ports, analog-to-digital converters, pulse-width modulators, serial ports, and specialized communication devices.

Microsecond one millionth of a second.

Millisecond one thousandth of a second.

Modem an acronym for modulator-demodulator used for phone-line communications between terminals and remote computers, later between two or more computers. (Digi International refers to some of its modules as modems, which they are, but because modem sounds anachronistic, I identify the Digi products as XBee modules.)

Node an end point in a wired or wireless network or a junction between components in a schematic diagram or circuit.

Ohm the unit used to define the resistance of an electrical component. Named after Georg Ohm (1789–1854), a German physicist.

Op-amp see Operational Amplifier

Operational amplifier a versatile integrated-circuit amplifier that uses external components such as resistors and capacitors to tailor its function to specific analog-signal applications. An op-amp circuit can add two voltages, amplify two voltages, operate as part of a filter circuit, and so on.

Packet a "package" of information that conforms to a specific structure that governs its arrangement of data, the use of a checksum, and other characteristics.

PAN see Personal Area Network

Parity a system that helps communication equipment and memories detect errors in information by including an extra bit. An even-parity arrangement appends a bit to a series of bits so the result has an even number of 1 bits. Say you want to send the byte 11000111 with an even parity bit, as agreed with the recipient in advance. To have an even number of 1's in the data, you must append a 1 to the communication: 110001111. The parity bit, shown in boldface type, now gives an even number of 1's in the data. If you sent the data 00110011 with even parity, you would append a zero and transmit: 001100110 because the data already has an even number of 1's. Odd parity works the same way, but it ensures an odd number of 1's in a transmission.

Personal area network (PAN) a network with only a few devices under your control.

Port usually electrical connections on a computer that communicate with a specific type of device. Computers have serial ports, Ethernet ports, USB ports, and so on. Engineers often call a group of signals an I/O port.

Pot see Potentiometer

Potentiometer a device that lets a person control a resistance that varies from 0 ohms to *x* ohms, as specified by the potentiometer manufacturer. Small potentiometers, often called a trimmer, pot, or trim pot, let technicians or service people make slight adjustments to a resistance.

Pull-up part of a circuit that "pulls up" a pin or connection to a voltage, usually through a resistor.

Pulse-width modulation (PWM) also pulse-width modulator; a technique that varies the width of a continuous train of electrical pulses that occur at a preset frequency. Pulse widths can vary from 0 to 100 percent. The PWM technique lets a device such as a microcontroller create a semi-analog output that, with a low-pass filter, becomes a true analog voltage.

PWM see Pulse-Width Modulation

Resistor an element that impedes the flow of electricity.

Serial port a connection to a UART, or UART-like device, for asynchronous-serial communications. See also Universal Asynchronous Receiver Transmitter.

Start bit the first bit, a logic 0, in a transmission from a UART.

Stop bit the last bit, a logic 1, in a transmission from a UART.

Trigger an electronic signal that initiates an event.

Trimmer see Potentiometer

UART see Universal Asynchronous Receiver Transmitter

Universal asynchronous receiver transmitter (UART) usually hardware designed specifically for serial communications at standard bit rates. A UART transmits and receives information in a standard format with equally timed bits. A logic-0 start bit begins each transmission, followed by five to as many as eight data bits. An optional parity bit can follow the data bits. At the end of the data, a UART sends one or two stop bits (logic 1). The transmitting and receiving UARTs must have the same settings for the transmission rate, the number of data bits (usually eight), the type of parity (usually none), and the number of stop bits (usually one). The bits appear at a UART one right after the other without any "space" or "dead time" between them. The output uses a non-return-to-zero format, which means no signal transition occurs between consecutive 1's or 0's in a transmission. Infrequently programmers implement a UART in software.

Universal serial bus (USB) a standardized bus for high-speed communications between a computer and an external device such as a DVD player, printer, or electronic instrument. A USB connection operates at several standard data rates and USB cables employ several types of standard connections.

USB see Universal Serial Bus

Volt the standard unit for the measurement of electrical potential, named for Alessandro Volta (1745–1827), perhaps best known for the for the invention of the battery.

XBee a wireless type of module manufactures by Digi International.

X-CTU a Windows program created by Digi International that simplifies communication between a PC and XBee modules as well as the display of communicated information, testing and configuring XBee devices.

ACKNOWLEDGEMENT

Thanks go to Wikipedia for information about Ampère, Ohm, and Volta.

Index

Printed in the United States
By Bookmasters